TIMAIOS OF LOCRI,
ON THE NATURE OF THE WORLD AND THE SOUL

Society of Biblical Literature

TEXTS AND TRANSLATIONS
GRAECO-ROMAN RELIGION SERIES

edited by
Hans Dieter Betz
Edward N. O'Neil

Texts and Translations 26
Graeco-Roman Religion Series 8

TIMAIOS OF LOCRI,
*ON THE NATURE OF THE WORLD
AND THE SOUL*
Text, Translation, and Notes by
Thomas H. Tobin

TIMAIOS OF LOCRI,
ON THE NATURE OF THE WORLD AND THE SOUL

Text, Translation, and Notes
by
Thomas H. Tobin

Scholars Press
Chico, California

TIMAIOS OF LOCRI,
ON THE NATURE OF THE WORLD AND THE SOUL

Text, Translation, and Notes
by
Thomas H. Tobin

© 1985
The Society of Biblical Literature

Greek text adapted from Timaeus Locrus, *De natura mundi et animae*, ed. and trans. by W. Marg (Philosophia Antiqua 24; Leiden: Brill, 1972). Used with permission of E. J. Brill.

Library of Congress Cataloging in Publication Data

Peri psychas kosmō kai physios. English.
  On the nature of the world and the soul.

  (Texts and translations ; 26) (Graeco-Roman religion series ; 8)
  Translation of: Peri psychas kosmō kai physios.
  Bibliography: p.
  Includes index.
  1. Cosmology, Ancient. 2. Soul—Early works to 1800. 3. Mind and body—Early works to 1850. I. Timaeus, of Locri. II. Tobin, Thomas H., 1945- . III. Title. IV. Title: Timaios of Locri, On the nature of the world and the soul. V. Series. VI. Series: Graeco-Roman religion series ; 8.
BD495.P4713   1984   113   84-5498
ISBN 0-89130-767-2
ISBN 0-89130-742-7 (pbk.)

Printed in the United States of America
on acid-free paper

TABLE OF CONTENTS

	Page
PREFACE	vii
INTRODUCTION	1
The Date of the TL	3
The TL and Plato's *Timaeus*	7
The Literary Character of the TL	17
The Manuscripts of the TL	20
NOTES TO THE INTRODUCTION	23
SELECT BIBLIOGRAPHY	29
OUTLINE OF THE TREATISE	31
TEXT AND TRANSLATION	32
NOTES TO THE TRANSLATION	75
INDEX VERBORUM	81

PREFACE

My interest in Timaeus of Locri's *On the Nature of the World and the Soul* began as part of my research into the Middle Platonic influences on the interpretations of the biblical accounts of the creation of man by Philo of Alexandria. I found that the last translation of this treatise into English had been in the middle of the nineteenth century (Bohn's Classical Library). At the suggestion of Professor John Strugnell of Harvard University, I decided to make the treatise more readily available to English-speaking students of ancient religion. This project was made much easier by the existence of a very fine critical edition of the treatise and German translation by Professor Walter Marg and by a first-rate commentary on it by Professor Matthias Baltes. I am much indebted to the work of both of these scholars. I want to express my gratitude to Professors Hans Dieter Betz of the University of Chicago and Edward N. O'Neil of the University of Southern California, the editors of this series, for their careful editing and many helpful suggestions. I also want to thank Rev. Theodore J. Tracy, S.J., Professor Emeritus of Classics of the University of Illinois (Circle Campus) for his careful reading of and helpful comments on the translation of the treatise. Finally, thanks are due to Loyola University of Chicago for granting me a sabbatical semester, part of which was spent on the completion of this translation.

Thomas H. Tobin, S.J.
Loyola University of Chicago
January, 1984

INTRODUCTION

Middle Platonism is the particular form that the Platonic philosophical tradition took between roughly 80 B.C. and A.D. 220. It represented a rejection of the scepticism of the Middle and New Academies (ca. 280-80 B.C.) and a renewed interest in Platonic cosmology. In that form, the Platonic philosophical tradition is of great interest to students of ancient religion. This is because of its immense influence on the religious outlooks and sensibilities of the early centuries of the Christian era. During this period it was probably the single most important philosophical tradition in the Graeco-Roman world. It also deeply affected early Christian writers such as Clement of Alexandria (ca. A.D. 150-215) and Origen (ca. A.D. 185-254) to name but two. Middle Platonism also influenced the Jewish biblical interpreter, Philo of Alexandria (ca. 20 B.C.-A.D. 50), and through him much of early Christian interpretation of the Bible. Through the discovery of the Gnostic manuscripts at Nag Hammadi, we are becoming more aware of the extent to which Middle Platonism influenced various Gnostic patterns of thought. Among these Gnostic manuscripts found at Nag Hammadi, one even finds a Coptic translation of a small section of Plato's *Republic* (CG VI,5).

While the influence of Middle Platonism was pervasive, our knowledge of it remains remarkably fragmentary. Very little of Middle Platonism comes to us directly. Most of what we know of it is derived from what other writers say about Middle Platonists. We have very few complete works from Middle Platonic authors themselves. From the middle of the second century A.D. we have extant works of Albinus (fl. A.D. 150) and Apuleius of Madaura (ca. A.D. 123-180). From the first century A.D. or the early second century A.D. we have treatises of Plutarch of Chaeronea (ca. A.D.

50-120) and the present work, Timaeus of Locri's *On the Nature of the World and the Soul* (TL).

The value of this treatise, however, goes beyond the fact that it is one of the few extant Middle Platonic philosophical works. Because Middle Platonism's renewed interest in cosmology centered on Plato's *Timaeus* and its interpretation, we have in the TL one of the few complete examples of how this crucial text of Middle Platonism was understood and interpreted. As we shall see, the interpretation of the *Timaeus* found in the TL was probably not very original. But, for that very reason, it gives us a view of an "ordinary" Middle Platonic interpretation of Plato's *Timaeus*. This is important because Middle Platonism was so influential on the various religious traditions of the period primarily through its reinterpretations of Plato's cosmology. The present treatise is important, then, not only for students of ancient philosophy but also for students of early Christianity, Hellenistic Judaism, and Gnosticism.

Timaeus of Locri's *On the Nature of the World and the Soul* purports to be the original words of the central character of Plato's *Timaeus*.[1] Since Timaeus was from Locri, a town in southern Italy, the TL was written in Doric, the Greek dialect of that area. While it does differ in significant ways from what Timaeus says in Plato's dialogue, the TL contains basically the same material found in *Timaeus* 27c-92c: the basic causes of the universe, the ordering of the body of the world, the introduction of the world-soul, and the fashioning of the human body and soul. Often the parallels between the two are very close. The TL is not, however, a dialogue nor does it contain any of the preliminary discussions found in Plato's *Timaeus* (17a-27b). The TL, then, appears to be the work of a fifth-century Greek writer, Timaeus of Locri, a work which was then used by Plato about sixty years later as the basis for his dialogue.

Introduction

The TL, however, is not attested before the second century A.D. It was first mentioned by the Neopythagorean Nicomachus of Gerasa (fl. A.D. 100) and the Middle Platonist Calvenus Taurus (fl. A.D. 145).[2] Both accepted the work as genuine. This is not surprising since, from at least the third century B.C. on, Plato had been accused of composing his dialogue on the basis of a book by an unnamed Pythagorean, which he had purchased at a high price.[3] It seemed quite natural to assume that the book that Plato had purchased and then used was the work of Timaeus of Locri.

The Date of the TL

Today there is general agreement that the true relationship between the TL and Plato's *Timaeus* is precisely the reverse. The TL depends on Plato's dialogue. Despite the significant differences between the two, much of the TL reads like an epitome of Plato's *Timaeus*—and a somewhat bland epitome at that. Had a native of Locri written the TL, one would expect his native Doric dialect to have been used consistently. But such is not the case. One even finds Koine forms used along with the Doric forms of the *same* word.[4] Finally, and most importantly, there are numerous post-Platonic elements in the TL, the most obvious of which is the use of the Aristotelian concept of matter (ὕλη).[5]

While there is a consensus that the document is post-Platonic, the question of exactly how post-Platonic it is remains the focus of debate. There is a firm *terminus ante quem*. Since the TL was known to both Nicomachus of Gerasa and Calvenus Taurus, it must have been composed before the end of the first century A.D. The question is where the document is to be placed between Plato and Nicomachus. In order to answer that question, one must look for details in the text which can be dated by reference to other authors and works.

The earliest date has been suggested by Gilbert Ryle, the Oxford linguistic analyst.[6] According to Ryle, the TL was composed by Aristotle in 361 B.C. as a gift to Dionysius of Syracuse. At the time of its composition, Plato was still alive and Aristotle was still a young man.[7] Ryle argues that the non-Platonic elements of vocabulary in the document are Aristotelian and, often enough, are found only in Aristotle and the TL.[8] Ryle also claims that at thirteen points the TL improves on Plato's *Timaeus* and that none of these improvements "presupposes any post-Aristotelian speculations or discoveries; and nearly all are made by Aristotle, often in early works."[9] For example, TL 55 says that cold blocks up the pores. Plato says nothing about the pores, but Aristotle (*De caelo* 307b) mentions that cold can block up the pores.[10] Finally, Ryle claims that, with the exception of one or two doubtful cases, "no Stoic words, logical, philosophical, or scientific" are found in the treatise.[11] Aristotelian influences on the content and vocabulary of the treatise are indeed considerable, but Ryle fails to notice that some of the vocabulary in the treatise is probably post-Aristotelian.[12] We shall see later that there are additional elements in the TL that suggest a date several centuries later.

The other scholar to assign the TL a relatively early date is Holger Thesleff.[13] Following R. Harder, Thesleff distinguishes two levels of the TL. The first level (Q) encompasses most of the document; the second level (F) includes a few short, often inconsistent additions and the fiction that the document was by Timaeus of Locri.[14] Since he sees the TL as one of a number of Pythagorean pseudepigrapha of the Hellenistic period, Thesleff dates the first level (Q) of the TL to ca. 300 B.C. and the second level (F) to ca. 200 B.C.[15] More specifically, Thesleff argues that the post-Platonic elements in the TL are derived from the Old Academy of Crantor, Speusippus, and Xenocrates.[16] In response to criticism, especially that of W. Burkert,

Introduction   5

which makes such an early date impossible, Thesleff now
suggests a date toward the end of the second century B.C.
or very early in the first century B.C.[17]

However, it is doubtful that all these Pythagorean
pseudepigrapha should be grouped together, and it is even
more doubtful that the TL should be grouped with them.[18]
Unlike Pythagorean pseudepigrapha, the TL is a summary and
interpretation of a well-known text central to the *Platonic*
tradition. It more likely represents developments of the
Platonic tradition rather than developments in
Pythagoreanism.

Most scholars date the TL to the first century B.C. or
or the first century A.D. Harder gives the first century
B.C. for most of the document (Q) but the first century
A.D. for the actual forgery itself (F).[19] Anton, Taylor,
Cornford, Baltes, and Dillon prefer the *late* first century
B.C. or the first century A.D.[20] M. Baltes has provided
the most detailed explanation for this dating in his re-
cent commentary on the TL.[21]

According to Baltes, elements in the TL place it
within the thought world of Middle Platonism, that is,
within the forms that the Platonic tradition took from ca.
80 B.C. to ca. A.D. 220.[22] For example, TL 7 states that
"before the heaven came to be, the idea and matter, as
well as the god who is the fashioner of the better already
existed." The basic components of the universe, then, are
three: the god, the idea, and matter. That position,
according to Baltes, indicates that the TL emerged in
Middle Platonism after Antiochus of Ascalon (ca. 130-68
B.C.).[23] Baltes also points out that the contrast between
the ideal world and the visible world found in TL 43 first
appears in Philo of Alexandria (ca. 20 B.C.-A.D. 50) (*Op.*
15,24). The same is true for the use of the term εἰκών
(image) in the sense of "model" rather than "copy" (TL 43,
88); it too first appears in Philo (*Som.* 1.79).[24]

Other elements in the TL have led Baltes to suggest an even more specific location, that is, in the circle of another Alexandrian, the Middle Platonist Eudorus (fl. 20 B.C.). He lists five details in which the TL agrees with what little we know of the figure of Eudorus of Alexandria.[25]

1. Both Eudorus and the TL, following Crantor (ca. 335-275 B.C.), take the number 384 as the basis for the division of the world-soul.[26]

2. Eudorus defines "passion" in a Stoic fashion as an "excessive impulse" (πλεονάζουσα ὁρμή); the author of the TL seems to have had that same definition in mind when he describes the passions as "impulses" (ὁρμαί).[27]

3. Both the TL and Eudorus show a great interest in mathematics, especially as it affects the division of the world-soul.[28] This interest in the mathematical structure of the world may reflect the influence of Pythagorean thought on both Eudorus and the TL.

4. Following Xenocrates (ca. 356-314 B.C.), both use the terms "male" and "female" in their explanations of basic causes of the world.[29]

5. Both Eudorus and the author of the TL deny that the description of the ordering of the world is meant to be a temporal description.[30]

While admitting that these points of agreement between the TL and Eudorus are not conclusive, Baltes claims that they give a certain probability to the notion that this document comes from the circle of Eudorus.[31] None of the points of coincidence between Eudorus and the TL, however, are shared by them alone. Even the two most important points of contact between Eudorus and the TL (the number 384 as the basis for the division of the world-soul, and the use of the terms "male" and "female" in explaining the basic elements of the universe) go back to the Old Academy, to Crantor and Xenocrates respectively. In addition, the TL does not contain any of the central philosophical positions associated with Eudorus (i.e. a supreme One, a lesser One, and an Unlimited Dyad).[32]

Yet there may be some cumulative effect to the number of coincidences between Eudorus and the TL. One can probably say that the TL came after Eudorus and that the author of the TL was aware of his work.[33] It seems more likely that the author of the TL was aware of these details of the thought of Eudorus than that each drew separately, for example, on two different members of the Old Academy. Whatever one's position on this particular issue, it is fairly clear that the TL should be placed within the Middle Platonic thought world of the late first century B.C. or the first century A.D. The details that it shares with the works of both Philo of Alexandria and Eudorus of Alexandria also suggest that the TL may have originated in Alexandria.

The TL and Plato's *Timaeus*

Since the TL is a Middle Platonic reformulation of Plato's *Timaeus*, it is best understood in comparison with Plato's dialogue. While the TL closely follows the *Timaeus* at many points, there are significant differences between the two, and these differences illuminate the character and significance of the TL. Some of these differences are matters of approach, others of content, and still others of structure.

1. The TL differs from the *Timaeus* in approach in several significant ways.

(a) Plato's *Timaeus* is a dialogue between Socrates and his three friends, Timaeus, Hermocrates, and Critias. As the dialogue progresses, Timaeus gradually becomes the dominant figure. His descriptions of the ordering of the universe are quite long, and after *Tim*. 29d, Timaeus is the only speaker for the rest of the dialogue.

The TL, however, makes no pretense of being a dialogue. It begins with "Timaeus of Locri said the following...." Because it is no longer a dialogue, the TL contains none of the preliminary conversations between Socrates and his friends (*Tim*. 17a-27b). For example, the TL does not contain Critias' story of the lost island of Atlantis (*Tim*. 21b-26a).

(b) In the *Timaeus*, Plato, through the character of Timaeus, is quite conscious of the problematic nature of his attempt to explain the origin and structure of the universe. The visible universe is always changing, never stable; it is always in a state of becoming. Because of this instability, Timaeus admits that it is impossible "to render an account at all points entirely consistent with itself and exact" (*Tim*. 29c).[34] The best that can be hoped for is that one can give an explanation that is a likely story or account (εἰκὼς μῦθος or λόγος), an account no less likely than any other (*Tim*. 29c-d). The characterization of Timaeus' description of the origin and structure of the universe as "likely" appears twenty-seven times in Plato's dialogue. This strong sense of methodological reserve is absent from the TL. The description in the TL appears to be a straightforward, almost dogmatic description of the origin and structure of the universe. It begins without any methodological preface but immediately launches into a description of the basic elements of the universe (TL 1-7). There is no sense of the difficulty, even the impossibility of the task, of describing the origins of the universe.

(c) Since Plato's Timaeus thought that his description could be only a "likely account," he allows himself a certain freedom in the use of language. His language sometimes has a mythological quality to it. For example, Timaeus describes the mixing of human souls as taking place in "the same mixing bowl wherein he (the Demiurge) had mixed and blended the soul of the universe" (*Tim*. 41d). Again, in connection with the creation of human souls, the Demiurge directly addressed the traditional Greek gods (Zeus, Hera, etc.) whom he had just made and told them that he would make the divine and ruling part of the human soul but that they would be responsible for the rest (*Tim*. 41a-d).

Both the metaphor of the "mixing bowl" and the speech to the traditional gods are missing from the parallel sections of the TL (43-45). In fact, the traditional gods play no role whatsoever in the fashioning of human beings. That role is now played by a blander and non-mythological "changeable nature" (ἡ ἀλλοιωτικὰ φύσις) (TL 44). Because the author of the TL does not view this description of the origin and structure of the world as only a "likely account" but as a straightforward description, as no longer metaphorical but literal, whatever smacks of the mythological in Plato's *Timaeus* has been weeded out. Paradoxically, the lack of methodological reserve has led to a type of demythologizing.

(d) Finally, while the Timaeus of Plato's dialogue does not present his description of the world in any strictly logical fashion, nevertheless he gives rather detailed explanations of why the universe is constructed in a certain way. This occurs most clearly at the beginning of his discourse. He explains why one must distinguish between *being* and *becoming* and between the *intelligible* and the *sensible* and why *becoming* and the *sensible* are only pale reflections of *being* and the *intelligible* (*Tim*. 27d-29d). There are other places in his description where he also gives a fairly lengthy explanation of why something has to be the way that it is.[35]

Reasons why some aspect of the world is the way it is are also found in the TL.[36] But they are far less elaborate, and they are essentially statements rather than explanations. For example, the motive for the god's ordering of the world is given in one short sentence: the god wanted to put order into matter (TL 7). In Plato's *Timaeus*, the explanation for the ordering of the world extends over several paragraphs (*Tim*. 29d-30c). The same is true when one compares the explanations of why the body of the world needs no limbs (*Tim*. 33b-34a; TL 17). The TL states what Plato's Timaeus tries to explain.

The approaches, then, of the two documents are very different. While Plato's *Timaeus* maintains a methodologically sophisticated even playful attitude toward the problems involved, the TL is earnest, didactic, and even pedantic in its approach.

2. Other differences between Plato's *Timaeus* and the TL are less matters of approach than they are of content. For the most part, these changes involve either the omission, the addition, or the clarification of details from Plato's *Timaeus*.[37]

(a) Omissions:

(1) The diagram of the heavens (*Tim.* 36b-d).

(2) The fashioning of the heavenly gods (*Tim.* 39e-40b).

(3) The different grades of purity of the elemental bodies (*Tim.* 41d).

(4) Physiological and medical details (*Tim.* 44d-45b; 45d-46a; 75e-76e; 82b; 83a-e; 85a-b; 90e-92c).

(b) Additions:

(1) Explanation of why the earth is the oldest element (TL 31).

(2) Explanation of the dodecahedron as the image of the universe (TL 35).

(3) A detailed description of the divisions of the world-soul (TL 21-23).

(4) A discussion of the problem of the morning and evening star (TL 27-28).

(c) Clarifications and alterations:

(1) Clearer description of the movement of the planets (*Tim.* 38c-39d; TL 26-29).

(2) Clarification of the different types of elemental bodies and their construction (cube, pyramid, octahedron, icosahedron, and dodecahedron) (*Tim.* 53c-56c; TL 33-36).

(3) Clarification of the proportions that bind the universe together (*Tim.* 31b-32c; TL 39-41).

(4) Refinements and alterations of physiological and medical details (TL 52/*Tim.* 61c-64a, expansion of the role of touch; TL 53-54/*Tim.* 62c-64a, relationship of light and heavy, above and below; TL 57/*Tim.* 66d, the sense of smell;

TL 58/*Tim*. 67a-c, the phenomenon of sound; TL 60/*Tim*. 77c-e, the role of *pneuma*; TL 62/*Tim*. 77c-e, no void in the body; TL 64/*Tim*. 79c-e, the phases of breathing).[38]

The reasons for these omissions, additions, and clarifications or alterations are varied. The omissions are, at least partially, due to the summary character of the TL. But, as Baltes points out, other omissions, as well as the additions and clarifications, are due to the author's attempt to update or modernize Plato's *Timaeus* in the light of developments in Hellenistic astronomy and medicine.[39] These details, of course, also serve to point to a later rather than an earlier date for the TL.

3. Finally, there are structural differences between Plato's *Timaeus* and the TL. At several crucial points the author of the TL has rearranged the material from Plato's *Timaeus*. These rearrangements give us additional clues about how Plato's *Timaeus* was understood and interpreted in Middle Platonic circles.

The main part of Plato's dialogue, Timaeus' discourse (*Tim.* 27c-92c), following some preliminary discussion (*Tim.* 27c-29d), falls into three main sections. The first (*Tim.* 29d-47e) describes the ordering of the universe through the craftsmanship of mind (νοῦς) (*Tim.* 47c). In this section there are descriptions of the Demiurge, the Demiurge's use of intelligible models, the fashioning of the body of the world, the world-soul, and finally the human soul and body. All of these are described from the point of view of the ordering power of mind.

The second section (*Tim.* 47e-69a) describes roughly the same process, only now from the point of view of necessity (ἀνάγκη). Necessity is at the other end of the spectrum from mind. Timaeus calls it the "errant cause" (ἡ πλανωμένη αἰτία) (*Tim.* 48a). It is that aspect of reality in which things occur of themselves (in that sense "necessary") but without further purpose (in that sense random and without order). Necessity is that aspect of reality which is permanently recalcitrant to the ordering power of mind.[40] Timaeus then goes on to describe the notions of space and chaos, the construction of the elemental bodies, and the various types of sense perception.

In the third and final section of Timaeus' discourse (*Tim.* 69a-92c), he discusses the cooperation and interaction of mind and necessity. Here he describes the mortal parts of the soul, the structure of the human body, the diseases of the body and the soul, and their cures.

The threefold movement of thought in Plato's *Timaeus* means that the world is looked at from three points of view, and that the same reality is looked at from the top down (mind), from the bottom up (necessity), and from the middle where the two interact. This is most obvious in Plato's treatment of human beings. The structure of human beings is treated at length in each of the three sections (*Tim.* 41d-47e; 61c-68d; 69d-92c). The movement of thought in the *Timaeus* is circular in the sense that Plato returns three times to the same realities, but each time from a different point of view.

The author of the TL, however, by the rearrangement of these three major blocks of material has significantly altered the structure of Plato's *Timaeus*.[41]

(1) The material in *Tim.* 39e-45b, which primarily describes the structure of the human body and soul, has been moved down so that in the TL it occurs after the description of the elemental bodies (TL 43-46).

(2) The material in *Tim.* 69d-76e, which deals with the mortal parts of the soul and the basic structure of the human body, has been moved up in the TL so that it is now placed just before the section on the various types of sense perception (TL 46-47).

(3) The material in *Tim.* 47e-53c, which primarily looks at what results from necessity, space, and chaos, has been moved to the very beginning of the TL and brought together with a description of the basic works of mind (TL 1-7).[42]

Introduction

The result of these rearrangements is that all of the material that involves cosmology has been grouped together (TL 1-41), while all of the material that treats anthropology has likewise been grouped together (TL 43-86). The TL, then, unlike Plato's *Timaeus*, has only two major sections, the one concerned with the order and structure of the universe and the other concerned with human beings and other mortal creatures.

This restructuring of the *Timaeus* is obviously a simplification of the more complicated patterns of the *Timaeus*. Yet it is more than that. The TL is also a Middle Platonic interpretation of Plato's *Timaeus*, an interpretation accomplished by rewriting.

One result of this restructuring is that the order in which topics are treated reflects an order common among Middle Platonists. In the material which is parallel to that found in the TL, both Albinus' *Didaskalikos* (second century A.D.) and Apuleius' *De Platone et eius dogmate* (second century A.D.) follow the same basic order which is found in the TL. Both Albinus and Apuleius first deal with cosmology and then with anthropology.[43] Cosmology and anthropology are not mixed together as they are in Plato's *Timaeus*. One doubts that such a rearrangement of material originated with the author of the TL. Rather, all three writers (the author of the TL, Albinus, and Apuleius) reflect a standard Middle Platonic way of interpreting Plato's *Timaeus*.[44]

A second result of this rearrangement, and perhaps the reason for it, is that it allows the author of the TL to interpret Plato's views on the basic principles or causes of the universe in characteristically Middle Platonic terms. This emerges when one examines the section in the TL that deals with first principles (TL 1-7), a section that combines *Tim.* 29d-31b and *Tim.* 47e-53c.

The TL opens with a description of the two basic causes of everything: mind and necessity (TL 1). Mind is

the principle of the best things, necessity of what is
limited by the properties of the elemental bodies. These
are the two basic causes described in Plato's *Timaeus*
(47e-48e). In addition, mind is called a god, an identification not made explicitly in the *Timaeus*, but quite a
natural conclusion given that mind is described by Plato
as the *divine* cause (τὸ θεῖον) (*Tim.* 68e). However, the
rest of the description of the basic principles of the
universe (TL 2-7) differs substantially from that found in
Plato's *Timaeus*. The differences are most apparent when
one reads the summary of TL 2-6 found in TL 7: "According
to this account then, before the heaven came to be, the
idea and matter, as well as the god who is the fashioner of
the better, already existed." The basic causes of the
universe are no longer two (mind and necessity) but three
(the god, the idea, and matter).[45] Three significant
shifts of interpretation have taken place in which (1)
necessity (ἀνάγκα) has been identified with Aristotle's
prime matter (ὕλη), (2) a triad of mind, idea, and matter
has emerged, and (3) the idea has taken on greater
importance.

The identification of necessity with matter is not
found in Plato; the process seems to have begun with Aristotle. Aristotle (*Phys.* 209b,11-17) identified Plato's
notion of space with his own concept of matter. That identification became common among Middle Platonists.[46] Then,
because Plato did not clearly distinguish necessity from
space, some Middle Platonists also identified these two.[47]
This meant that it was easy for the author of the TL to
identify all three, necessity, matter, and space (TL 4, 7,
32).

The second shift of interpretation in the TL is the
emergence of a triadic structure for the basic causes of
the universe: mind or the god, the idea, and matter.
Plato's *Timaeus* has a more complex and less clear description of the basic components of the universe, components

Introduction   15

that cannot be reduced to a triad.[48] But this triadic
structure of mind, the ideas, and matter is common in
Middle Platonism. It first appears in Varro (116-27 B.C.)
and is also found in Plutarch, Albinus, Apuleius, and
Aetius.[49] In fact, some scholars think that this triadic
structure is the most common characteristic of Middle Platonism.[50] In this scheme, the ideas are often thoughts in
the mind of the god who fashions the universe.[51] In some
Middle Platonists, however, the triad of mind, the ideas,
and matter is part of a larger metaphysical scheme. In
this larger scheme, the god or mind that fashions the universe is not the primal or highest deity but an intermediate figure between the primal deity and matter. The triad
of the god or mind that fashions the universe, the ideas,
and matter is often part of a larger triad of a primal
deity, an intermediate figure who fashions the universe,
and matter.[52] The ideas now become thoughts in the mind
of the intermediate figure. Such, however, is not the case
in the TL. There is no indication in the TL that there is
any more basic structure to reality than mind, the idea,
and matter. There is no larger metaphysical triad of which
this smaller triad is a part. This larger metaphysical
triad does not appear in Varro but does appear in both
Albinus and Apuleius. In this respect, the TL represents
a less developed, less elaborate form of Middle Platonism
than does Albinus or Apuleius.[53]

The third shift of interpretation in the TL is the
growth of the importance of the idea. In the TL the term
ἰδέα (idea) is used only in the singular, never in the plural. Plato himself did use "idea" in the singular (e.g.
Plt. 263b) but only seldom; for the most part he used the
plural "ideas." The same is true of the term εἶδος (form)
when it is used as a synonym for "idea."[54] The closest the
TL comes to a plural usage is in TL 11, which speaks of the
idea as the "model, containing in itself all intelligible
living beings." Yet even here the idea is thought of in

the singular (παράδειγμα), which then contains the intelligible living beings.[55] This is in contrast to Varro and to Middle Platonists such as Albinus, Apuleius, and the position mentioned by Aetius. All of these have the triad of mind, the *ideas*, and matter; they look on the ideas as a plurality.[56] The "idea," however, in the TL has become more of an intermediate *figure* between mind and matter. It is more than a collection of individual ideas or even a realm of ideas. One impetus for such a development may have been the notion that the ideal world, unlike either matter or the sensible world, is indivisible (ἀμέριστος) (TL 3, 18). In a sense, the idea in the TL holds a position that is analogous to the intermediate figure of the larger metaphysical schemes one finds in Albinus and Apuleius. Along with mind and matter, the idea is one of the three substances that form the basis for the rest of the universe (TL 3-7).

The importance given to the idea in the TL, while unusual in Middle Platonism, is not unique. The position taken by the author of the TL, when he speaks of the idea in the singular and as an intermediate figure, is similar to that found in Philo of Alexandria. In *Op.* 24-25 the intelligible world (ὁ νοητὸς κόσμος), that is, the world of ideas, is identified with the intermediate figure of the *Logos*. As in the TL, so too in Philo, the world of ideas has coalesced into a unified, intermediate figure, the *Logos*.[57]

The structural similarity of Philo and the TL on this issue once again suggests that the TL is connected with the kind of Middle Platonism one finds in Alexandria at the end of the first century B.C. or in the first century A.D. In addition, one can see from the differences between the TL and Plato's *Timaeus*, that the Middle Platonism of the TL has also incorporated elements of the Peripatetic (e.g. matter replacing necessity), Stoic (e.g. passion as an impulse), and Pythagorean (e.g. the mathematical divisions of

Introduction

the world-soul) philosophical traditions.[58] This again is characteristic of Middle Platonism.

The Literary Character of the TL

Turning to the style and the pseudepigraphical character of the TL, one is struck by the unevenness of the work. At times the TL seems to be a summary of what is found in Plato's *Timaeus*. Material which had been scattered throughout Plato's *Timaeus* is often brought together and summarized thematically by the author of the TL. On the other hand, sections of the TL are reinterpretations of of Plato's *Timaeus*, some of which are substantial. Finally, the attempt to put the discourse in the mouth of Timaeus of Locri is superficial. The Doric dialect is not carried through consistently, nor is the direct discourse one would expect to follow the opening phrase (Timaeus of Locri said the following: "...") consistently maintained. As a matter of fact, much of the TL is in *indirect* discourse.[59]

The uneven quality of the TL has led Harder and Baltes to suggest that there are several levels. Harder posits two levels in the TL. The first is a Hellenistic revision of Plato's *Timaeus* made in the second, or more probably the first, century B.C. (Q).[60] This makes up most of the document. It is a commentary which flattens out, dogmatizes, and simplifies Plato's *Timaeus*. While basically Platonic, especially in its interest in mathematics, some Peripatetic elements have been introduced.[61] Finally, Harder suggests that the author of Q was also interested in medical questions.[62] The second level is that of the forger (F) who puts the TL into the mouth of Timaeus of Locri. Harder dates this level to the first century A.D. The Doric dialect of the document as well as a good deal of its style (or lack of it) also belong to this level. According to Harder, the one responsible for this level was particularly interested in mathematics and astronomy. He also attributes to the forger the inconsistencies and whatever else seemed to him muddled and unclear in the text.[63]

Baltes sees the TL basically as a combination of an epitome of Plato's *Timaeus* and a commentary on the *Timaeus*. All of the material that summarizes (with some rearrangement) Plato's dialogue belongs to the epitome. All of the material that alters or adds to positions taken in the *Timaeus* comes from the commentary on the *Timaeus*. In Baltes' view, the commentary is not a serious work, and he prefers to see it as lecture notes taken by a student, perhaps of the school of Eudorus of Alexandria.[64] There may have been a third hand involved, that of the forger. Baltes is uncertain, however, whether or not the forger is to be identified with the one responsible for the lecture notes.[65]

While it is true that the text is uneven, that some sections of the TL are simply summaries of Plato's *Timaeus* while others are reinterpretations or additions, and that some sections are less clear than others, it is still far from certain that the text of the TL is composite. Prescinding from the questions of the Doric dialect and the forgery for a moment, the uneven quality may result from a combination of the nature of the text, that is, an interpretative rewriting of Plato's *Timaeus*, and the inadequacies of the author. Most Middle Platonists accepted some parts of the *Timaeus* and altered and reinterpreted other parts. If that is the case, one expects some unevenness. In addition, no one would claim philosophical genius for the author of the TL. The TL probably reflects opinions or interpretations initiated by others. Once again, this makes for unevenness in the text. There is, then, no compelling reason to think of the TL as a document composed from several earlier works.

While the author of the TL was hardly a great philosopher, he was not incompetent. It is a bit more likely,

Introduction

therefore, that the one responsible for the Doric dialect and for putting the treatise into the mouth of Timaeus of Locri was someone other than the author of the treatise itself. The Doric dialect in the TL does remain superficial, and the use of direct and indirect discourse is very inconsistent. Both the literary fiction and the Doric dialect may be the work of someone else. One suspects that the TL was originally not pseudonymous but simply anonymous. Placing it in the mouth of Timaeus of Locri was a natural enough step since it was a reformulation of Timaeus' discourse in Plato's dialogue. That step may have been more a a literary convention than a serious attempt to deceive the reader. After all, there was a story that Plato had purchased a book on which he based his dialogue.[66] That story may have served as one impetus for putting the TL into the mouth of Timaeus of Locri. Another may have been the existence by the early first century B.C. of a collection of Pseudopythagorean writings.[67] Imitation of these writings may also have played a role in turning the TL into a pseudepigraphical work. Ironically, by the second century A.D., the TL came to be accepted as genuine, and that acceptance no doubt was an important factor in its preservation by and influence on later interpreters of Plato.[68]

Basically, then, the TL is a reformulation of Plato's *Timaeus* in Middle Platonic thought patterns. Since it reads like an epitome even when it reinterprets Plato, one doubts that the author of the TL is an original writer. Rather, the TL represents a didactic summary of already existing Middle Platonic interpretations and modernizations of the *Timaeus*. Its natural setting is a Middle Platonic school of some sort, perhaps in Alexandria at the end of the first century B.C. or in the first century A.D.[69] Its significance lies less in its philosophical analysis than in the opportunity it gives its readers to see more clearly into Middle Platonism at the end of the first century B.C. and in the first century A.D., the formative period of

Middle Platonism. One can see a bit more clearly *how* Middle Platonists arrived at their characteristic philosophical positions from reading and interpreting Plato's *Timaeus*. One can also see how the patterns of Middle Platonic thought represented by the TL influenced someone like Philo of Alexandria and, through Philo, an important segment of early Christian thought.

The Manuscripts of the TL

W. Marg, for his critical edition of the TL, was able to locate and collate fifty-two manuscripts of the TL.[70] While most of the manuscripts are from the fourteenth and fifteenth centuries, the oldest manuscripts come from the thirteenth century and the beginning of the fourteenth century. Since Marg was able to establish a stemma, only a few of the manuscripts are important for establishing the text.

N	Neopolitanus Bibl. Naz. 312 (A.D. 1314)
B	Parisinus graecus 1808 (13th century). This manuscript is missing TL 78 (ἁρμόξατο) to the end.
A	Biblioteca Angelica 107 (late 13th century or 14th century)
L	Laurentianus Plut. cod. 59,1 (14th century)
Par. 1809	Parisinus graecus 1809 (15th century)
E	Biblioteca Escorialensis cod. gr. 306 (*olim* y I 13) (late 13th century or early 14th century). This manuscript is missing TL 53 (προκρίνει) to the end.
W	Vindobonensis suppl. philos. graec. 7 (15th century)
V	Marcianus f.a. 185 (14th century)

Of these, manuscripts N, B, and E are the principal ones. Manuscripts A, L. and Par. 1809 are used for the missing section of B; and W and V are used for the missing section of E. Manuscript N stands apart, while B and E are more closely related and are derived from the same ancestor.

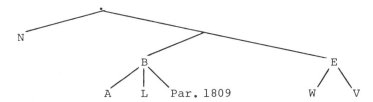

A special textual problem is found in TL 22-23. These two paragraphs give an explanation of the mathematical divisions of the world-soul mentioned in TL 21. This explanation, which is partially based on musical progressions, is found in manuscripts N, E, W, V, and L. In N it is found in the text itself. In the other manuscripts (E, W, V, and L), the explanation is found in the margins or at the end. Marg thinks that this explanation comes from manuscript N and is introduced into the other manuscripts. From a text-critical viewpoint, the explanation is secondary in manuscripts other than N.

The question is whether the mathematical explanation found in N was originally part of the text of the TL or whether it was added later to clarify the numbers found in TL 21 (the division of the world-soul). Both Marg and Baltes doubt that the mathematical explanation found in TL 22-23, in its present form, was originally a part of the TL. The fullness of detail in TL 22-23 is stylistically unlike the conciseness of the rest of the TL. In addition, there is little attempt in TL 22-23 to maintain the Doric dialect of the rest of the document.[71]

While the mathematical explanation in its present form is secondary, some explanation must have stood there originally. TL 19, as well as passages from Nicomachus and Proclus, all point in that direction.[72] What may have stood there originally was either a simple mathematical table of progressions or a table with only very short explanatory remarks.[73] Since the explanation found in N is secondary,

I have included in this translation instead a version suggested both by Marg and Baltes of what the original mathematical table looked like.[74]

1. The table is meant to fill out the mathematical progressions between the numbers found in *Tim.* 35b-c: 1, 2, 3, 4, 8, 9, 27. These numbers are the unit (1), the first even number (2), the first odd number (3), the square of 2 (4), the cube of 2 (8), the square of 3 (9), and the cube of 3 (27).

2. The number 384 was chosen as the base because it is the first whole number on which all of the progressions involved result in *whole* numbers.

3. Starting with the base 384, the intervals between the progressions 1, 2, 3, 4, 8, 9, 27 are filled in with the appropriate whole tones (an interval of 9/8) and two kinds of semitones, the *leimma* (an interval of 256/243) and the *apotome* (an interval of 2187/2048). The two types of semitones are indicated by l. (*leimma*) and ap. (*apotome*) in the table.

4. There are 36 members in the progression, and their sum total is 114,695.

The text used in this translation is that of Marg. I have also made use of suggestions from Baltes (Timaios Lokros) and from T. Szlezák's review of Marg.[75] These emendations are noted in the apparatus.

NOTES TO THE INTRODUCTION

[1] The correct title, *On the Nature of the World and the Soul*, is given by Iamblichus (*In Nic.* 105.10).

[2] Nicomachus of Gerasa, *Harm.* 11; Calvenus Taurus *apud* Joannes Philoponus, *De aet. mun. contra Proc.* 6.8.

[3] Timon of Phlius (ca. 320-230 B.C.) *apud* Aulus Gellius, *N. A.* 3.17.4. For other versions of this type of story, see W. Burkert, *Lore and Science in Ancient Pythagoreanism* (Cambridge: Harvard University, 1972) 223-27.

[4] For example: νοατός (3)--νοητός (10); δυνάμιες (68) --δυνάμεις (1); ἔασσα (33)--οὖσα (21). For a detailed treatment of the dialect, see Timaios Lokros, *Über die Natur des Kosmos und der Seele*, commented on by M. Baltes (Philosophia Antiqua 21; Leiden: Brill, 1972) 11-19. The inconsistencies are too numerous to be attributed to the vagaries of textual transmission.

[5] See especially TL 2-7,32. Other post-Platonic elements will be mentioned in the course of this introduction.

[6] G. Ryle, "The Timaeus Locrus," *Phronesis* 10 (1965) 174-90.

[7] Ibid., 185-86. The reasons for the specific date are less important than the claim that the TL was composed by Aristotle.

[8] Ibid., 175-76.

[9] Ibid., 178.

[10] Ibid., 177.

[11] Ibid., 176.

[12] For example: ποτιγειότατος (26), ἀνισόπλευρον (33), ἰσονομία (41), ἀντίλαψις (48), διαστατικός (55), ἐναέριος (60), σύρροος (62), ἀντεπεισάγω (64), ἁπλαῖ δυνάμιες (68), δυσαισθησία (71). Other examples are found in the commentaries of J.R.W. Anton, *De origine libelli* "ΠΕΡΙ ΨΥΧΑΣ ΚΟΣΜΩ ΚΑΙ ΦΥΣΙΟΣ" *inscripti qui vulgo Timaeo Locro tribuitur* (Part 1.1, Erfurt: Carl Villaret, 1883; Part 1.2, Numburg: Alb. Schirmer, 1891) and Baltes, Timaios Lokros. It is obviously impossible to say when most words were first used. But the fact that so many words in the TL are attested only after Aristotle is significant and counts against Aristotle's authorship.

[13] H. Thesleff, *An Introduction to the Pythagorean Writings of the Hellenistic Period* (Acta Academiae Aboensis, Humaniora 24.3; Åbo: Åbo Akademi, 1961) 59-65, 102.

[14] Ibid., 60-62. See pp. 17-18 of this introduction for a fuller description of Harder's position.

[15] Ibid., 62-65, 102.

[16] Ibid., 61: 384 as the basic number for the divisions of the world-soul (TL 21) (Crantor); the 36 members (ὅροι) of the division of the world-soul (TL 21) (Crantor); the terms ἡμιτρίγωνον and ἡμιτετράγωνον (TL 33) (Speusippus); ἀρχαί as male and female (TL 5) (Xenocrates).

[17] H. Thesleff, "On the Problem of the Doric Pseudo-Pythagorica. An Alternative Theory of Date and Purpose," in *Pseudepigrapha I: Pseudopythagorica--Lettres de Platon--Littérature pseudépigraphique juive* (Entretiens sur l'Antiquité classique 18; Geneva: Fondation Hardt, 1972) 83-84. W. Burkert's criticisms are found in two articles: "Zur geistesgeschichtlichen Einordnung einiger Pseudopythagorica," in *Pseudepigrapha I*, 25-55; review of *The Pythagorean Texts of the Hellenistic Period*, edited by H. Thesleff, *Gnomon* 39 (1967) 548-56. In his review article, Burkert points out that the use of the term μοῖρα (TL 29) to mean one of the 360 degrees of a circle first appears in Hypsicles of Alexandria (fl. 150 B.C.) (p. 555). This means that the TL must have been written after the middle of the second century B.C. at the earliest.

[18] See Burkert, "Zur geistesgeschichtlichen Einordnung einiger Pseudopythagorica," 25-55.

[19] R. Harder, "Timaios[4]," PW 6A (1936) 1226.

[20] Anton, *De origine libelli*, 1.1.31 (first century A.D. at the earliest); A. E. Taylor, *A Commentary on Plato's Timaeus* (Oxford: Clarendon, 1928) 656-57 (early first century A.D.); F. M. Cornford, *Plato's Cosmology: The Timaeus of Plato* (1937; rpt. Indianapolis: Bobbs-Merrill, 1957) 3 (first century A.D.); Baltes, Timaios Lokros, 25 (first century A.D. or very late first century B.C.); J. Dillon, *The Middle Platonists* (Ithaca: Cornell University, 1977) 131 (first century A.D.).

[21] Baltes, Timaios Lokros, 20-26.

[22] For a description of the basis elements of Middle Platonism, see Dillon, *The Middle Platonists*, 43-51.

[23] Baltes, Timaios Lokros, 47-48.

[24] Baltes, Timaios Lokros, 21. Baltes claims that this meaning of εἰκών may have originated with the circle of Antiochus of Ascalon. That is far less certain, although this meaning of the term probably did not originate with Philo.

[25] Ibid., 23. For a discussion of Eudorus, see H. Dörrie, "Der Platoniker Eudorus von Alexandreia," in *Platonica Minora* (Munich: Wilhelm Fink, 1976) 297-309; Dillon, *The Middle Platonists*, 115-35.

[26] TL 21; Eudorus *apud* Plutarch, *De proc. an. in Tim.* 16, 1020C.

[27] TL 74; Eudorus *apud* Stobaeus, *Ecl.* 2.44.5.

[28] TL 21; Eudorus *apud* Plutarch, *De proc. an. in Tim.* 16, 1019E.

[29] TL 5; Eudorus *apud* Simplicius, *In phys.* 181.25-26.

[30] TL 7; Eudorus *apud* Plutarch, *De proc. an. in Tim.* 3, 1013A-B. The position held by Eudorus and the TL is the more common opinion among ancient interpreters of Plato's *Timaeus*.

[31] Baltes, Timaios Likros, 23.

[32] See Eudorus *apud* Simplicius, *In phys.* 181.7-30.

[33] See Dillon, *The Middle Platonists*, 131. This is also the opinion of Taylor (*Commentary on Plato's Timaeus*, 656) and W. Theiler ("Philo von Alexandria und der hellenisierte Timaeus," in *Philomathes: Studies and Essays in the Humanities in Memory of Philip Merlan*, ed. R. B. Palmer and R. Hamerton-Kelly [The Hague: Nijhoff, 1971] 25).

[34] The translations of Plato's *Timaeus* are from Cornford, *Plato's Cosmology*.

[35] For example: the motive of creation (*Tim.* 29d-30c); why there is only one world (*Tim.* 31a-b); why there are four types of elemental bodies (*Tim.* 31b-32c); why the body of the world is without limbs (*Tim.* 33b-34a); why sight is so important (*Tim.* 47b-c); the reason for the ideal models of the four elemental bodies (*Tim.* 51b-e); the reason for two basic types of triangle in the construction of the world (*Tim.* 54a-b). This list could be easily expanded.

[36] TL 4, 7, 8, 9, 10, 16, 17, 20, 27, 30, 31, 34, 35, 52-53, 54, 62, 85.

[37] This list is taken from Baltes, Timaios Lokros, 7-10.

[38] The specific reasons for these changes are given by Baltes (Timaios Lokros) at the appropriate points in his commentary.

[39] Baltes, Timaios Lokros, 9-10.

[40] See Cornford, *Plato's Cosmology*, 159-77.

[41] Harder ("Timaios^4," 1205-20) provides a chart which compares the arrangement of the TL with that of Plato's *Timaeus*.

[42] There are also several minor shifts, but these follow in the wake of the three major shifts of material. *Tim.* 31b-32c, 34c-d, 42a-d, 47a-c, 80a have all been rearranged in the TL as a result of the three major shifts.

[43] Cosmology: Albinus, *Didaskalikos* 8-16, 162.21-172.16; Apuleius, *De dog. Plat.* 1.5-12. Anthropology: Albinus, *Didaskalikos* 17-26, 172.17-179.29; Apuleius, *De dog. Plat.* 1.13-18.

[44] There is a distinct possibility that part of Albinus' *Didaskalikos* is a revised version of a work by the late first-century B.C. Alexandrian philosopher Arius Didymus. If that is the case, then some of the material in the *Didaskalikos* may be from the same milieu as the TL. See Dillon, *The Middle Platonists*, 269, 286-87, 290, 304.

[45] TL 7 indicates that matter has been identified with necessity and is not simply a derivative of it. TL 32 points in the same direction. For another interpretation, see Baltes, Timaios Lokros, 34.

[46] See Plutarch, *De proc. an. in Tim.* 24, 1024C; *De Is. et Os.* 53, 372F; 56, 373F.

[47] See Plutarch, *De proc. an. in Tim.* 6, 1014E-F; Numenius, Fr. 52, 299 (des Places).

[48] This is especially true for the notions of necessity, the receptacle, and chaos; see Cornford, *Plato's Cosmology*, 160-210.

[49] Varro *apud* Augustine, *De civ. dei* 7.28; Plutarch, *Quaest. conviv.* 8.2.4, 720A-C; Albinus, *Didaskalikos* 9, 163.10-164.5; Apuleius, *De dog. Plat.* 1.5-6; Aetius 1.3.21.

[50] See H. Dörrie, "Ammonios, der Lehrer Plotins," in *Platonica Minora*, 342; W. Theiler, *Die Vorbereitung des Neuplatonismus* (Berlin: Weidmann, 1930) 18-19.

[51] Aetius 1.3.21; Albinus, *Didaskalikos* 9, 163.27-28. The same is also probably true for Varro (*apud* Augustine, *De civ. dei* 7.28).

Notes to the Introduction

[52] Albinus, *Didaskalikos* 10, 164.16-165.5; Apuleius, *De dog. Plat.* 1.6.

[53] This is a matter of the structure of thought, not a matter of chronology. If Albinus' *Didaskalikos* is a revised version of the work of Arius Didymus, then the TL and sections, at least, of Albinus come from roughly the same time period. See n. 44.

[54] ἰδέα: 2, 3, 6, 7(2), 10, 86; εἶδος: 5, 32, 88.

[55] In TL 86, the term "idea" is used in the sense of "shape" or, perhaps, "semblance." The use in TL 86 is non-technical and does not enter into the present discussion since it does not refer to the question of Platonic "ideas" or "archetypes."

[56] See n. 49.

[57] The *Logos* in the writings of Philo is a complex figure. What Philo says of the figure may represent several generations of thought. *Op.* 24-25 represents that point in the development of thought where the ideas have been brought together to form an intermediate figure.

[58] Given the fact that Timaeus Locri was a "Pythagorean," there is surprisingly little (beyond the mathematical divisions of the world-soul) which is clearly Pythagorean. As both Taylor (*Commentary on Plato's Timaeus*, 660, 662, 663) and Harder ("Timaios[4]," 1222, 1224) have pointed out, Pythagorean elements in the TL are superficial.

[59] See Anton, *De origine libelli*, 597-99.

[60] Harder, "Timaios[4]," 1223-25. The second-century B.C. date is too early.

[61] Ibid., 1224. For example: the identification of "below" and "middle," "circumference" and "above" (TL 54; Aristotle, *Phys.* 191a); the analogy of the virtues of the body and the soul (TL 79; Stobaeus, *Ecl.* 124-26).

[62] Harder, "Timaios[4]," 1225.

[63] Ibid., 1220-23, 1226.

[64] Baltes, Timaios Lokros, 24-25.

[65] Ibid., 25-26.

[66] See n. 3.

[67] See Alexander Polyhistor (1st cent. B.C.) *apud* Diogenes Laertius 8.24; Thesleff, "On the Problem of the Doric Pseudo-Pythagorica," 80.

[68] See n. 2.

[69] Baltes (Timaios Lokros, 4) calls it a "Lehrschrift."

[70] For a detailed description of all of the manuscripts and their relationships to one another, see Timaeus Locrus, *De natura mundi et animae*, ed. and trans. by W. Marg (Philosophia Antiqua 24; Leiden: Brill, 1972) 1-52.

[71] Ibid., 60-66.

[72] Nicomachus of Gerasa, *Harm.* 11; Proclus, *In Tim.* 3.188.

[73] A mathematical table (Baltes, Timaios Lokros, 80-81); a mathematical table with short explanatory remarks (Marg, ed., *De natura mundi et animae*, 74).

[74] For more detailed explanations of the table, see Timaeus the Locrian, *The Soul of the World and Nature*, trans. by G. Burges (Bohn's Classical Library, The Works of Plato, Vol. 6; London: Bohn, 1854) 6.169-73; Baltes, Timaios Lokros, 79-82; Marg, ed., *De natura mundi et animae*, 67-75.

[75] T. A. Szlezák, review of Timaios Lokros (*Über die Natur des Kosmos und der Seele*) and Timaeus Locrus (*De natura mundi et animae*), *Gnomon* 48 (1976) 135-44.

SELECT BIBLIOGRAPHY

Editions and Translations

The Pythagorean Texts of the Hellenistic Period. Ed. by H. Thesleff. Acta Academiae Aboensis, Humaniora 30.1. Åbo: Åbo Akademi,, 1965, pp. 202-25.

Timaeus Locrus. *De natura mundi et animae.* Ed. and trans. by. W. Marg. Philosophia Antiqua 24. Leiden: Brill, 1972.

Timaeus the Locrian. *The Soul of the World and Nature.* Trans. by G. Burges. Bohn's Classical Library. The Works of Plato. Vol. 6. London: Bohn, 1854, pp. 145-73.

Secondary Literature

Anton, J.R.W. *De origine libelli.* "ΠΕΡΙ ΨΥΧΑΣ ΚΟΣΜΩ ΚΑΙ ΦΥΣΙΟΣ" *inscripti qui vulgo Timaeo Locro tribuitur.* Part 1.1, Erfurt: Carl Villaret, 1883; Part 1.2, Numburg: Alb. Schirmer, 1891.

Burkert, W. *Lore and Science in Ancient Pythagoreanism.* Cambridge: Harvard University, 1972.

_____. "Zur geistesgeschichtlichen Einordnung einiger Pseudopythagorica." In *Pseudepigrapha I: Pseudopythagorica--Lettres de Platon--Littérature pseudépigraphique juive.* Entretiens sur l'Antiquité classique 18. Geneva: Fondation Hardt, 1972, pp. 23-55.

_____. Review of *The Pythagorean Texts of the Hellenistic Period.* Ed. by H. Thesleff. *Gnomon* 39 (1967) 548-56.

Cornford, F. M. *Plato's Cosmology: The Timaeus of Plato.* Indianapolis: Bobbs-Merrill, 1957 (orig. 1937).

Dillon, J. *The Middle Platonists.* Ithaca: Cornell University, 1977.

Harder, R. "Timaios^4." PW 6A (1936) cols. 1203-26.

O'Daly, G.J.P. Review of Timaios Lokros, *Über die Natur des Kosmos und der Seele,* commented on by M. Baltes; and, Timaeus Locrus, *De natura mundi et animae,* ed. and trans. by W. Marg. *The Classical Review* 25 (1975) 197-99.

Ryle, G. "The Timaeus Locrus." *Phronesis* 10 (1965) 174-90.

Szlezák, T. A. Review of Timaios Lokros, *Über die Natur des Kosmos und der Seele*, commented on by M. Baltes; and Timaeus Locrus, *De natura mundi et animae*, ed. and trans. by W. Marg. *Gnomon* 48 (1976) 135-44.

Taylor, A. E. *A Commentary on Plato's Timaeus*. Oxford: Clarendon, 1928.

Thesleff, H. *An Introduction to the Pythagorean Writings of the Hellenistic Period*. Acta Academiae Aboensis, Humaniora 24.3. Åbo: Åbo Akademi, 1961.

──────. "On the Problem of the Doric Pseudo-Pythagorica. An Alternative Theory of Date and Purpose." In *Pseudepigrapha I: Pseudopythagorica--Lettres de Platon--Littérature pseudépigraphique juive*. Entretiens sur l'Antiquité classique 18. Geneva: Fondation Hardt, 1972, pp. 57-87.

Timaios Lokros. *Über die Natur des Kosmos und der Seele*. Commented on by M. Baltes. Philosophia Antiqua 21. Leiden: Brill, 1972.

OUTLINE OF THE TREATISE

I. Cosmology: The Order and Structure of the World (1-42)

- 1-6   Principles and Causes (*Tim.* 27c-31b; 48e-52d)
- 7-17   The Ordering of the Body of the World (*Tim.* 28a-34b; 52d-53b)
- 18-21   The Ordering of the World-Soul (*Tim.* 34b-36e)
- (22-23   Detailed Description of the Divisions of the World-Soul)
- 24-29   Astronomy: The Stars and the Planets (*Tim.* 36b-e; 38c-39e; 40a-d)
- 30   Time (*Tim.* 37c-38c)
- 31   The Earth (*Tim.* 40b-c)
- 32-42   The Elemental Bodies (*Tim.* 31b-32c; 53c-61c)

II. Anthropology: Human Beings and Other Mortal Creatures (43-86)

- 43-47   The Fashioning of Human Beings and Other Mortal Creatures (*Tim.* 39e-45b; 69d-74e)
- 48-59   Types of Sense Perception (*Tim.* 47a-e; 61c-68d; 80a)
- 60-67   Nutrition and Breathing (*Tim.* 70a-b; 77c-81e; 84d)
- 68-70   Diseases of the Body (*Tim.* 81e-86a)
- 71-77   Diseases of the Soul (*Tim.* 42a; 69c; 86b-87b; 91a)
- 78-86   Cures for the Diseases of Both the Body and the Soul (*Tim.* 87b-91a)

III. Conclusion (87-88) (*Tim.* 92c)

Τιμαίω Λοκρῶ
Περὶ φύσιος κόσμω καὶ ψυχᾶς

205   1. Τίμαιος ὁ Λοκρὸς τάδε ἔφα·   93a
    Δύο αἰτίας εἶμεν τῶν συμπάντων, νόον
μὲν τῶν κατὰ λόγον γιγνομένων, ἀνάγκαν δὲ
τῶν βίᾳ καττὰς [δυνάμεις τῶν σωμάτων.]
τουτέων δὲ τὸ μὲν τᾶς τἀγαθῶ φύσιος εἶμεν
θεόν τε ὀνυμαίνεσθαι ἀρχάν τε τῶν ἀρίστων·
τὰ δ' ἑπόμενά τε καὶ συναίτια ὄντα ἐς
ἀνάγκαν ἀνάγεσθαι. 2. τὰ δὲ ξύμπαντα τρία·   b
ἰδέαν, ὕλαν, αἰσθητὸν τὸ οἷον ἔγγονον
τουτέων. 3. καὶ τὰν μὲν εἶμεν ἀεί, ἀγέ-
νατόν τε καὶ ἀκίνατον, ἀμέριστόν τε καὶ
τᾶς ταὐτῶ φύσιος, νοατάν τε καὶ παράδειγμα
τῶν γεννωμένων, ὁκόσα ἐν μεταβολᾷ ἐντι·
τοιοῦτον γάρ τι τὰν ἰδέαν λέγεσθαί τε καὶ   94a
νοῆσθαι. 4. τὰν δ' ὕλαν ἐκμαγεῖον καὶ
ματέρα τιθάναν τε καὶ γεννατικὰν εἶμεν τᾶς
τρίτας οὐσίας· δεξαμέναν γὰρ τὰ ὁμοιώματα
206 ἐς αὐτὰν καὶ οἷον ἐναπομαξαμέναν ἀποτελῆν
τάδε τὰ γεννάματα. ταύταν δὲ τὰν ὕλαν
ἀίδιον μὲν ἔφα, οὐ μὰν ἀκίνατον, ἄμορφον
δὲ κατ' αὐταύταν καὶ ἀσχημάτιστον, δεχομέναν
δὲ πᾶσαν μορφάν· τὰν δὲ περὶ τὰ σώματα
μεριστὰν εἶμεν καὶ τᾶς θατέρω φύσιος.
ποταγορεύοντι δὲ τὰν ὕλαν τόπον καὶ χώραν.   b
5. δύο ὧν αἴδε ἀρχαί ἐντι, ἇν τὸ μὲν εἶδος
λόγον ἔχει ἄρρενός

205   1. Timaeus of Locri spoke as follows: 93a

There are two causes of all things: Mind, of what comes into being according to due proportion,[1] and Necessity, of what comes into being under constraint in accordance with [the characteristic powers of the elemental bodies.] Of these causes, one is of the nature of the good, and is called a god and the source of the best things. The others, however, which are secondary and contributory causes, are attributed to Necessity. 2. All things   93b
are made up of these three: idea, matter, and the sensible, which is, as it were, the offspring of the other two. 3. The idea is eternal, ungenerated, and immovable, indivisible and of the nature of the Same, intelligible and the model of generated things, which are in a state of change. The idea is spoken and thought of as some-   94a
thing of that sort. 4. Matter, however, is a recipient of impressions and mother, nurse and generative of the third substance. For, after it has received the likenesses in itself and, as it were, stamped itself, with them, matter produces those things
206   which are generated. He said that this matter was everlasting, but not immovable, in itself patternless and shapeless, but receiving every pattern. Because of its relationship to the elemental bodies, it is divisible and of the nature of the Different. Matter is called place and   94b
space.[2] 5. These (two) then are principles, of which the form is analogous to male

τε καὶ πατρός, ἁ δ' ὕλα θήλεός τε καὶ ματέρος.
τρίτα δὲ εἶναι τὰ ἐκ τουτέων ἔγγονα. 6. τρία
δὲ ὄντα τρισὶ γνωρίζεσθαι, τὰν μὲν ἰδέαν νόῳ
κατ' ἐπιστάμαν, τὰν δ' ὕλαν λογισμῷ νόθῳ διὰ
τὸ μηδέπω κατ' εὐθυωρίαν νοῆσθαι ἀλλὰ κατ'
ἀναλογίαν, τὰ δ' ἀπογεννάματα αἰσθήσει καὶ
δόξᾳ.
    7. Πρὶν ὦν ὠρανὸν λόγῳ γενέσθαι ἤστην   94c
ἰδέα τε καὶ ὕλα καὶ ὁ θεὸς δαμιουργὸς τῶ
βελτίονος. ἐπεὶ δὲ τὸ πρεσβύτερον κάρρον
ἐστὶ τῶ νεωτέρω καὶ τὸ τεταγμένον πρὸ τῶ
ἀτάκτω, ἀγαθὸς ὢν ὁ θεὸς ὁρῶν τε τὰν ὕλαν
δεχομέναν τὰν ἰδέαν καὶ ἀλλοιουμέναν
παντοίως μέν, ἀτάκτως δέ, ἐδήλετο εἰς τάξιν
αὐτὰν ἄγεν καὶ ἐξ ἀοριστᾶν μεταβολᾶν ἐς
ὡρισμέναν καταστᾶσαι, ἵν' ὁμόλογοι ταὶ
διακρίσιες τῶν σωμάτων γίγνωνται, καὶ μὴ
κατ' αὐτόματον τροπὰς δέχηται.
    8. Ἐποίησεν ὦν τόνδε τὸν κόσμον ἐξ   94d
ἁπάσας τᾶς ὕλας, ὅρον αὐτὸν κατασκευάσας
τᾶς τῶ ὄντος φύσιος διὰ τὸ πάντα τὰ ἄλλα
ἐν αὐταύτῳ περιέχεν, ἕνα, μονογενῆ, τέλειον,
ἔμψυχόν τε καὶ λογικόν (κρέσσονα γὰρ τάδε
ἀψύχω καὶ ἀλόγω ἐστόν), καὶ σφαιροειδὲς
σῶμα (τελειότερον γὰρ τῶν ἄλλων σχημάτων
ἦν τοῦτο). 9. δηλόμενος ὦν ἄριστον
γέννημα ποιῆν,

---

8 ἤστην BE Proc. ποτ' ἤστην N.

[handwritten annotation: cf Plutarch: De Iside - 53 (372 F) 1315 is Matter, the cosmic receptacle]

and father, and the <u>matter is analogous
to female</u> and mother. The third is their
offspring. 6. Because they are three, they
are comprehended in three different ways:
the idea by the mind through knowledge;
matter by a spurious kind of reasoning,
because it can never be known directly
but only by analogy; and their progeny by
sense perception and opinion.

7. According to this account then,   94c
before the heaven came to be, the idea and
matter, as well as the god who is the
fashioner of the better, already existed.
Since the elder is superior to the younger
and the ordered is prior to the disordered,
the god, who is good and who saw matter
receiving the idea and being changed in all
kinds of ways but in a disordered manner,
wanted to put matter in order and to
bring it from a condition of indefinite
change into a state with a definite pattern
of change, so that the differences among
the elemental bodies might become propor-
tional and matter might not undergo alterations at random.

8. He made this world, then, from all   94d
of matter, constructing it as the limit of
the nature of being, since it contains
everything else in it; and he made it as
one, unique, and perfect, having both
soul and reason (for that is better than
being without soul and reason); and a
spherical body (for this is superior to
the other shapes).[3] 9. Therefore, as he
wanted to make the best creature, he made

τοῦτον ἐποίη θεὸν γεννατόν, οὔ ποκα φθαρη-
σούμενον ὑπ' ἄλλω αἰτίω ἔξω τῶ αὐτὸν
συντεταγμένω θεῶ, εἴ ποκα δήλοιτο αὐτὸν
διαλύειν· ἀλλ' οὐ γὰρ τἀγαθῶ ἐστιν ὁρμᾶν   94e
ἐπὶ φθορὰν γεννάματος καλλίστω· διαμένει
ἄρα τοιόσδε ὢν ἄφθαρτος καὶ ἀνώλεθρος καὶ
μακάριος. 10. κράτιστος δ' ἐστὶ γεννατῶν,
ἐπεὶ ὑπὸ τῶ κρατίστω αἰτίω ἐγένετο,
ἀφορῶντος οὐκ εἰς χειρόκματα παραδείγματα,
ἀλλ' ἐς τὰν ἰδέαν καὶ ἐς τὰν νοητὰν οὐσίαν,
ποθ' ἄνπερ τὸ γεννώμενον ἀπακριβωθὲν
κάλλιστόν τε καὶ ἀπαρεγχείρητον γίγνεται.
11. τέλειος δ' ἀεὶ κατὰ τὰ αἰσθητὰ ἐστιν,   95a
ὅτι καὶ τὸ παράδειγμα τῆνο αὐτῶ περιέχον
πάντα τὰ νοητὰ ζῷα ἐν αὐταύτῳ οὐδὲν ἐκτὸς
ἀπέλιπεν, ἀλλὰ ὅρος ἦν νοατῶν παντελής, ὡς
ὅδε ὁ κόσμος αἰσθητῶν.

12. Στερεὸς δὲ ὢν ἁπτός τε καὶ ὁρατὸς
γᾶς μεμοίρακται πυρός τε καὶ τῶν μεταξύ,
ἀέρος καὶ ὕδατος. 13. ἐκ παντελέων δὲ
συνέστακε σωμάτων, τάπερ ὅλα ἐν αὐτῷ ἐντί,
ὡς μή ποκα μέρος ἀπολειφθῆμεν ἐκτὸς αὐτῶ,   95b
ἵνα ᾖ αὐταρκέστατον τὸ τῶ παντὸς σῶμα
ἀκήρατόν τε τῶν ἐκτὸς κηρῶν· οὐ γὰρ ἦν τι
δίχα τουτέων· 14. ἀλλὰ καὶ τῶν ἐντός·
τὰ γὰρ καττὰν ἀρίσταν ἀναλογίαν συντεθέντα
ἐν ἰσοδυναμίᾳ

it a generated god, never to be destroyed
by any cause other than the god who had
given it order, if he would ever want to
dissolve it. But it is not the nature of   94e
the good to hasten the destruction of a
most beautiful creature. Because it is
such a creature, it continues to exist,
therefore, as incorruptible, indestructible,
and happy. 10. It is the best of creatures
because it came into being by means of the
best cause, a cause that looked not to
models made by hand but to the idea, that
is, to the intelligible substance; and
what is generated exactly from that same
substance is most beautiful and inviolable.[4]
11. It is forever perfect in the sensible   95a
realm because its model, containing in itself all intelligible living beings, left
nothing out but was an absolute limit in
the intelligible realm so that this world
is the limit in the sensible realm.

    12. Since the world is solid as well
as tangible and visible, it is composed of
earth and fire and the two intermediate
elemental bodies, air and water. 13. He
established it out of complete elemental
bodies; and these are totally within it,
lest any part be left outside of it, so   95b
that the body of the universe is completely
self-sufficient and unharmed by illness
coming from outside. For there was nothing
apart from these elemental bodies. 14. Nor
is it harmed by illness from within. For
elemental bodies put together according to
the best proportion in a balanced state,

οὔτε κρατεῖ ἀλλάλων ἐκ μέρεος οὔτε κρατέεται,
ὡς τὰ μὲν αὔξαν, τὰ δὲ φθίσιν λαμβάνεν, μένει
δ' ἐν συναρμογᾷ ἀδιαλύτῳ κατὰ λόγον ἄριστον.
15. τριῶν γὰρ ὡντινωνοῦν ὅρων ὅταν [καὶ]   95c
τὰ διαστάματα καττὸν αὐτὸν ἐστάθη λόγον ποτ'
ἄλλαλα, τότε δὴ τὸ μέσον ῥυσμῷ Δίκας ὁρήμεθα
ποττὸ πρᾶτον ὅ τί περ τὸ τρίτον ποτ' αὐτό,
καὶ πάλιν καὶ παραλλὰξ κατ' ἐφάρμοσιν τόπων
καὶ τάξιος. ταῦτα δ' ἀριθμῆμεν αἱ μὴ μετ'
ἰσοκρατίας ἀμάχανον παντί.

16. Εὖ δ' ἔχει καὶ καττὸ σχῆμα καὶ
καττὰν κίνασιν, καθ' ὃ μὲν σφαῖρα ὄν, ὡς
ὅμοιον αὐτὸ αὑτῷ παντᾷ εἶμεν καὶ πάντα   95d
τἆλλα ὁμογενέα σχάματα χωρῆν δύνασθαι,
καθ' ἂν δὲ ἐγκύκλιον μεταβολὰν ἀποδιδὸν
δι' αἰῶνος. μόνα δὲ ἁ σφαῖρα ἐδύνατο καὶ
ἀρεμέουσα καὶ κινευμένα ἐν τᾷ αὐτᾷ συναρ-
μόσθαι χώρᾳ, ὡς μὴ ποκα ἀπολεῖπεν μήτε
ἐπιλαμβάνεν ἄλλον τόπον, τῷ ἐκ μέσου ἴσον
εἶμεν παντᾷ. 17. λειότατον δ' ὂν ποτ'
ἀκρίβειαν καττὰν ἐκτὸς ἐπιφάνειαν οὐ
ποτιδέεται θνατῶν ὀργάνων, ἃ διὰ τὰς χρείας
τοῖς ἄλλοις ζώοις ποτάρτηταί τε καὶ διᾶκται.

18. Τὰν δὲ τῶ κόσμω ψυχὰν μεσόθεν   95e
ἐξάψας ἐπάγαγεν, ἔξω περικαλύψας αὐτὸν ὅλον
αὐτᾷ,

---

8 καὶ πάλιν coni. Baltes   κἂν πάλιν NBE Iambl.
κἄνπαλιν Thesleff et Marg   κἀνάπαλιν de Gelder.

neither overpower each other nor are they
overpowered so that some take on excess
and others deficiency; rather the world
remains in an indissoluble combination in
accordance with the best proportion.

15. Whenever the differences among any three terms are established according to the same proportion to each other, then we see that the middle term is related in the order of Justice to the first term in the same way as the third term is related to the middle term; the reverse and converse are also true, in keeping with adjustments of places and position.[5] To compute these things is altogether impossible except with a balance of powers.

16. The body of the universe is well off in terms both of shape and motion: in terms of shape, because it is a sphere and so is similar to itself at every point and can contain all other shapes of the same kind; and in terms of motion, because it eternally exhibits change in a circular movement. Only the sphere, both while at rest and in motion, was able to be fitted together into the same space so that it never gives up its place or takes another place, since it is at every point equidistant from the center. 17. Because completely and exactly smooth on its outer surface, it does not require those mortal organs which, because of need, are attached to and detached from other living beings.

18. He introduced the world-soul by fastening it in the middle, totally covering the world on the outside with it, and

κρᾶμα αὐτὰν κερασάμενος ἔκ τε τᾶς ἀμερίστω
μορφᾶς καὶ τᾶς μεριστᾶς οὐσίας, ὡς ἓν κρᾶμα
ἐκ δύο τουτέων εἶμεν. ᾧ ποτέμιξε δύο δυνάμιας   96a
ἀρχὰς κινασίων, τᾶς τε ταὐτῶ καὶ τᾶς τῶ
ἑτέρω· ἃ δὴ καὶ δύσμικτος ἔασσα οὐκ ἐκ τῶ
ῥᾴστω συνεκίρνατο. 19. λόγοι δ' οἵδε πάντες
ἐντὶ κατ' ἀριθμὼς ἁρμονικῶς συγκεκραμένοι.
ὣς λόγως κατὰ μοίρας διαιρήκει ποτ' ἐπιστάμαν,
ὡς μὴ ἀγνοῆν ἐξ ὧν ἁ ψυχὰ καὶ δι' ὧν συνεστάκει 20. (ἂν οὐχ ὑστέραν τᾶς σωματικᾶς
οὐσίας συνετάξατο ὁ θεός, ὥσπερ λέγομες ἁμές·
πρότερον γὰρ τὸ τιμιώτερον καὶ δυνάμει καὶ   96b
χρόνῳ· ἀλλὰ πρεσβυτέραν ἐποίη)· 21. μοῖραν
ἀφαιρέων τὰν πράταν μονάδων οὖσαν τεττόρων
ποτὶ ὀκτὼ δεκάσι καὶ τρισὶν ἑκατοντάσι.
ταύτας δὲ τάν τε διπλασίαν καὶ τριπλασίαν
ῥᾷον συλλογίξασθαι ἑσταμένω τῶ πράτω. δεῖ
δ' εἶμέν πως πάντας σὺν τοῖς συμπληρώμασι
καὶ τοῖς ἐπογδόοις ὅρους ἓξ καὶ τριάκοντα,
τὸν δὲ σύμπαντα ἀριθμὸν γενέσθαι μυριάδων
ια καὶ τετόρων χιλιάδων ἑξακατίων ϙε.   96c
22. Ταὶ δὲ διαιρέσιες αὗταί ἐντι·
[μυριάδες ‚ιαδ]

---

9 συνεστάκει B²E   συνέστακεν N
συνεστάκειαν B¹   συνέστακεν coni. Marg.

making it a mixture of the indivisible
pattern and the divisible substance, so that
one mixture arose from these two.[6] He also
blended with it two powers, the principles
of motion, that of the Same and that of the
Different. Because the power of the Different was difficult to blend, it was not
mixed easily. 19. All of these proportions
have been mixed together according to harmonious numbers. He also divided these
proportions into parts for the sake of
knowledge, lest anyone be ignorant of what
the soul was composed of and through what
means he established it. 20. (The god did
not really order the soul later than the
corporeal substance, as we seem to imply;
for the more valuable is prior in both power
and time; rather he made the soul earlier.)
21. He took one part as the first, a part
that consisted of four ones, plus eight tens
and three hundreds. From this one can
easily determine the square and the cube,
once the first number is established. In-
cluding the supplementary tones and the
wholetones, there must be altogether thirty-
six members; and the sum total must be
114,695.

22. The divisions are as follows:[7]

213 τὰν μὲν ὦν τῶ ὅλω ψυχὰν ταύτᾳ πως διεῖλε.

24. Θεὸν δὲ τὸν μὲν αἰώνιον νόος ὁρῆ μόνος, τῶν πάντων ἀρχαγὸν καὶ γενέτορα τουτέων. τῶν δὲ γεννατῶν ἕκαστον μὲν τᾷ ὄψει ὁρέομες, κόσμον δὲ τόνδε κατὰ μέρεα αὐτῶ ὁκόσα ὡράνιά ἐντι· τάπερ αἰθέρια ὄντα διαιρετὰ δίχα, ὡς τὰ μὲν τᾶς ταύτῶ φύσιος εἶμεν, τὰ δὲ τᾶς τῶ ἑτέρω. 25. ὧν τὰ μὲν ἔξωθεν ἄγει πάντα ἐν αὑτοῖς τὰ ἐντὸς ἀπ' ἀνατολᾶς ἐπὶ δύσιν τὰν καθ' ἅπαν κίνασιν, τὰ δὲ τᾶς τῶ ἑτέρω ἐντὸς ἀπὸ ἑσπέρας [τὰ] ποθ' ἕω μὲν ἐπαναφερόμενά τε καὶ καθ' αὑτὰ κινεόμενα, συμπεριδινέεται δὲ κατὰ συμβεβηκὸς τᾷ ταύτῶ φορᾷ, κράτος ἐχοίσᾳ ἐν κόσμῳ κάρρον.   96d

214 26. Ἃ δὲ τῶ ἑτέρω φορᾷ μεμερισμένα καθ' ἁρμονικῶς λόγως ἐς ἑπτὰ κύκλως συντέτακται. ἃ μὲν σελάνα ποτιγειοτάτα ἔασσα

---

10 ἅπαν NBE   ἁμέραν Hermann   ἁπλοῦν W.

Text and Translation   43

		(1)	(2)	(3)	(4)	(5)	(6)	(7)
(210)	1	384	432	486	512 1.	576	648	729
		(8)	(9)	(10)	(11)			
	2	768 1.	864	972	1024 1.			
		(12)	(13)	(14)				
(211)	3	1152	1296	1458				
		(15)	(16)	(17)	(18)	(19)	(20)	(21) (22)
	4	1536 1.	1728	1944	2048 1.	2187 ap.	2304	2592 2916
		(23)						
	8	3072 1.						
		(24)	(25)	(26)	(27)	(28)	(29)	(30) (31)
(212)	9	3456	3888	4374 1.	4608 1.	5184	5832	6144 1. 6561 ap.
						(32)	(33)	(34) (35)
						6912 1.	7776	8748 9216 1.
		(36)						
	27	10368						

213   In this way, then, he divided the soul of the whole.

24. Mind alone sees the eternal god, the originator of all things and their generator.[8] We see each generated being by means of sight; however, we <u>see</u> this world only in conjunction with those parts that are heavenly.[9] And they, being etherial, are divided into two types: some are of the nature of the Same, and others are of the nature of the Different. 25. Of these, the ones of the outer sphere guide all in the inner sphere, from sunrise to sunset, a motion which is that of the universe as a whole.[10] Those of the inner sphere, that is, of the Different, which go in the opposite direction, from evening to morning, and have their own proper motion, are whirled around, however, in dependence on the revolution of the Same, which has superior power in the world.

96d

26. The revolution of the Different,
214   which is divided according to harmonic ratios, is organized into seven orbits. Since the moon is closest to the earth, it finishes

ἔμμηνον τὰν περίοδον ἀποδίδωτι, ὁ δ' ἄλιος   96e
μετὰ ταύταν ἐνιαυσίῳ χρόνῳ τὸν αὐτῶ κύκλον
ἐκτελεῖ. δύο δ' ἰσόδρομοι ἀελίῳ ἐντί,
‘Ερμᾶ τε καὶ “Ηρας, τὸν ’Αφροδίτας καὶ
φωσφόρον τοὶ πολλοὶ καλέοντι. 27. νομῆς
γὰρ καὶ πᾶς ὅμιλος οὐ σοφὸς τὰ περὶ τὰν
ἱερὰν ἀστρονομίαν ἐστὶν οὐδ' ἐπιστάμων
ἀνατολᾶν τᾶν ἑσπεριᾶν καὶ ἑῴαν· ὁ γὰρ αὐτὸς
ποκὰ μὲν ἕσπερος γίγνεται, ἑπόμενος τῷ
ἀλίῳ τοσοῦτον ὁκόσον μὴ ὑπὸ τᾶς αὐγᾶς αὐτῶ
ἀφανισθῆμεν, ποκὰ δὲ ἑῷος, αἴκα προαγέηται   97a
τῶ ἀλίω καὶ προανατέλλῃ ποτ' ὄρθρον.
φωσφόρος ὢν πολλάκις μὲν γίγνεται ὁ τᾶς
’Αφροδίτας διὰ τὸ ὁμοδρομῆν ἀλίῳ, οὐκ αἰεὶ
δέ· ἀλλὰ πολλοὶ μὲν τῶν ἀπλανέων, πολλοὶ δὲ
τῶν πλαζομένων, πᾶς δ' ἐν μεγέθει ἀστὴρ ὑπὲρ
τὸν ὁρίζοντα πρὸ ἀλίου προγενόμενος ἀμέραν
ἀγγέλλει. 28. τοὶ δ' ἄλλοι τρεῖς, ῎Αρεός
τε καὶ Διὸς καὶ Κρόνω, ἔχοντι ἴδια τάχεα   97b
καὶ ἐνιαυτὼς ἀνίσως. ἐκτελέοντι δὲ τὸν
δρόμον περικαταλάψιας ποιεύμενοι φάσιάς τε
καὶ κρύψιας καὶ ἐκλείψιας γεννῶντες, ἀτρε-
κέας τε ἀνατολὰς καὶ δύσιας, ἔτι δὲ φάσιας
φανερὰς ἑῴας ἢ ἑσπερίας ἐκτελέοντι ποτὶ
τὸν ἄλιον· 29. ὃς ἀμέραν ἀποδίδωτι τὸν ἀπ'   97c
ἀνατολᾶς ἐπὶ δύσιν αὐτῶ δρόμον, νύκτα δὲ
τὸν ἀπὸ δύσεως ἐπ' ἀνατολάν· ἂν κίνασιν κατ'
ἄλλο ποιέεται, ἀγόμενος ὑπὸ τᾶς ταὐτῶ φορᾶς,
ἐνιαυτὸν δὲ

its circuit in a month; the sun which is next completes its orbit in a year. Two others have courses equal to that of the sun, the star of Mercury and the star of Hera, which people call the star of Venus and the Lightbringer.[11] 27. For shepherds and all ordinary people are not wise about what concerns sacred astronomy, nor do they understand the evening and morning risings. For the same star is now the evening star, when it follows the sun at such a distance that it is not hidden by the rays of the sun; and now the morning star, when it precedes the sun and, about dawn, rises before it. Therefore, the star of Venus is often the Lightbringer because it has the same course as the sun; but this is not always so. But many of the fixed stars, as well as many of the planets, in fact any heavenly body of a certain size when it comes over the horizon before the sun, announce the coming of the day. 28. The other three planets, those of Mars, Jupiter, and Saturn, have different speeds and unequal "years." The planets complete their courses, overtaking each other, and producing appearances, occultations, and eclipses, as well as precise risings and settings. They also complete their visible risings at morning or evening in relationship to the sun. 29. The sun brings about both day, the period from sunrise to sunet, and night, the period from sunset to sunrise. The sun completes this motion in dependence on another, that is, guided by the motion of the Same; but it completes the year in

96e

97a

97b

97c

καττὰν αὐτῷ καθ' ἑαυτὸν κίνασιν. ἐκ δὲ
τουτέων τῶν κινασίων, δύο ἐασσᾶν, τὰν ἕλικα
ἐκτυλίσσει, ποθέρπων μὲν κατὰ μίαν μοῖραν
ἐν ἀμερησίῳ χρόνῳ, περιδινεύμενος δὲ ὑπὸ
τᾶς τῶν ἀπλανέων σφαίρας καθ' ἑκάσταν
περίοδον ὄρφνας καὶ ἀμέρας.

30. Χρόνω δὲ μέρεα τάσδε τὰς περιόδως
λέγοντι, ὃν ἐγέννασεν ὁ θεὸς σὺν κόσμῳ.
οὐ γὰρ ἦν πρὸ κόσμω ἄστρα· διόπερ οὐδ'
ἐνιαυτὸς οὐδ' ὡρᾶν περίοδοι, αἷς μετρέεται
ὁ γεννατὸς χρόνος οὗτος. εἰκὼν δ' ἐστὶ τῶ
ἀγεννάτω χρόνω, ὃν αἰῶνα ποταγορεύομες·
ὡς γὰρ ποτ' ἀίδιον παράδειγμα, τὸν ἰδανικὸν
κόσμον, ὅδε ὡρανὸς ἐγεννάθη, [οὕτως] ὣς
ποτὶ παράδειγμα, τὸν αἰῶνα, ὅδε χρόνος σύν
κόσμῳ ἐδαμιουργήθη.

31. Γᾶ δ' ἐν μέσῳ ἱδρυμένα ἑστία θεῶν
ὥρός τε ὄρφνας καὶ ἀῶς γίνεται δύσιάς τε
καὶ ἀνατολὰς γεννῶσα κατ' ἀποτομὰς τῶν
ὁριζόντων, ὣς τᾷ ὄψει καὶ τᾷ ἀποτομᾷ τᾶς
γᾶς περιγραφόμεθα. πρεσβίστα δ' ἐντὶ τῶν
ἐντὸς ὡρανῶ σωμάτων· οὐδέ ποκα ὕδωρ ἐγεννάθη
δίχα γᾶς, οὐδὲ μάν τοι ἀὴρ χωρὶς ὑγρῶ, πῦρ
τε ἔρημον ὑγρῶ καὶ ὕλας ἃς ἐξάπτοι οὐκ ἂν
διαμένοι· ὥστε ῥίζα πάντων καὶ βάσις τῶν
ἄλλων ἁ γᾶ, καὶ ἐρήρεισται ἐπὶ τᾶς αὐτᾶς
ῥοπᾶς.

accord with its own proper motion.[12] As a result of these motions, two in number, the sun moves in a spiral, progressing one degree in the period of a day but spinning around under the influence of the fixed stars each period of night and day.

30. These revolutions are called the divisions of time, which the god generated along with the world. For there were no stars before the world. Therefore, there was neither the year nor the seasonal revolutions by which this generated time is measured. It is an image of ungenerated time, which we call eternity. For, just as this heaven was generated according to an everlasting model, the ideal world, so too this time was fashioned along with the world according to a model, that is, eternity.

31. Earth, which is set in the middle, is the hearth of the gods and the guardian of darkness and dawn and also produces settings and risings according to the segments of the horizons, for we define settings and risings by means of sight and the segment of the earth.[13] Earth is the oldest of the elemental bodies under heaven. Neither was water ever generated without earth, nor air without moisture; and fire deprived of moisture and matter, which it sets fire to, would not continue to exist. Thus, earth is the root of everything and the basis for everything else and is held firm by its own inclination toward the center.

32. Ἀρχαὶ μὲν ὦν τῶν γεννωμένων ὡς μὲν ὑποκείμενον ἁ ὕλα, ὡς δὲ λόγος μορφᾶς τὸ εἶδος· ἀπογεννάματα δὲ τουτέων ἐντὶ τὰ σώματα, γᾶ τε καὶ ὕδωρ ἀήρ τε καὶ πῦρ, ὧν ἁ γέννασις τοιαῦτα.

33. Ἅπαν σῶμα ἐξ ἐπιπέδων ἐντί, τοῦτο   98a
δὲ ἐκ τριγώνων, ὧν τὸ μὲν ὀρθογώνιον ἰσοσκελὲς ἡμιτετράγωνον, τὸ δὲ ἀνισόπλευρον, ἔχον τὰν μέζονα δυνάμει τριπλατίαν τᾶς ἐλάσσονος. ἁ δ' ἐλαχίστα ἐν αὐτῷ γωνία τρίτον ὀρθᾶς ἐστι, διπλασία δὲ ταύτας ἁ μέσα· δύο γὰρ τρίτων ἅδ' ἐστίν. ἁ δὲ μεγίστα ὀρθά, ἁμιόλιος μὲν τᾶς μέσας ἔασσα, τριπλατία δὲ τᾶς ἐλαχίστας. τοῦτο δ' ὦν τὸ τρίγωνον ἁμιτρίγωνόν   98b
ἐστιν ἰσοπλεύρῳ τριγώνῳ, δίχα τετμαμένῳ καθέτῳ ἀπὸ τᾶς κορυφᾶς ἐς τὰν βάσιν ἐς ἴσα μέρεα δύο. ὀρθογώνια μὲν ὦν ἐντι ἑκατέρω, ἀλλὰ ἐν ᾧ μὲν ταὶ δύο πλευραὶ ταὶ περὶ τὰν ὀρθὰν μόναι ἴσαι, ἐν ᾧ δὲ ταὶ τρεῖς πᾶσαι ἄνισοι. σκαληνὸν δὲ τοῦτο μὲν καλεέσθω, τῆνο δὲ ἁμιτετράγωνον, 34. ἀρχὰ συστάσιος γᾶς. τὸ γὰρ τετράγωνον ἐκ τουτέω, ἐκ τεττόρων ἡμιτετραγώνων συντεθειμένον· ἐκ δὲ τοῦ   98c
τετραγώνου γεννᾶσθαι τὸν κύβον, ἑδραιότατον καὶ σταδαῖον παντᾷ σῶμα, ἓξ μὲν πλευράς, ὀκτὼ δὲ γωνίας ἔχον.

32. The principles of generated things
are: matter as the substratum and form as the
basis of shape. The progeny of these are
the elemental bodies: earth and water, air
and fire. Their generation occurred in the
following way.

33. Every body is composed of surfaces   98a
and a surface is composed of triangles.
One of these triangles is an isosceles
right triangle which forms half of a square;
the other triangle does not have equal
sides, and the square of the longer side is
three times the square of the shorter side.
The smallest angle of this triangle is one-
third of a right angle. The middle one is
twice that size, that is, two-thirds of a
right angle. The largest is a right angle
which is half again as large as the middle
angle and three times as large as the
smallest angle. This triangle, then, is
half of an equilateral triangle which has   98b
been bisected perpendicularly from its
vertex to its base into two equal parts.
Each of these two types of triangles, therefore, is a right triangle; but in one the
two sides adjacent to the right angle are
equal, while in the other all three sides
are unequal. The latter is called a
scalene triangle and the former an isoceles
triangle.[14] 34. The isosceles triangle is
the principle of composition for earth.
The square comes from it, that is, it is put
together from four isosceles triangles.[15]
From the square comes the cube, the most   98c
stable and completely firm type of elemental
body. It has six sides and eight corners.

κατὰ τοῦτο δὲ βαρύτατόν τε καὶ δυσκίνατον ἁ
γᾶ, ἀμετάβλητόν τε σῶμα εἰς ἄλλα διὰ τὸ
ἀκοινώνατον εἶμεν τῶ ἄλλω γένεος τῶ τριγώνω·
μόνα γὰρ ἁ γᾶ ἴδιον στοιχεῖον ἔχει τὸ ἁμιτετράγωνον.

35. Τῆνο δὲ στοιχεῖον τῶν ἄλλων σωμάτων
ἐστί, πυρός, ἀέρος, ὕδατος. ἑξάκις γὰρ
συντεθέντος τῶ ἁμιτριγώνω τρίγωνον ἐξ αὐτῶ   98d
ἰσόπλευρον γίνεται· ἐξ οὗ ἁ πυραμὶς τέσσαρας
βάσιας καὶ τὰς ἴσας γωνίας ἔχοισα συντίθεται,
εἶδος πυρὸς εὐκινατότατον καὶ λεπτομερέστατον.
μετὰ δὲ τοῦτο ὀκτάεδρον, ὀκτὼ μὲν βάσιας,
ἓξ δὲ γωνίας ἔχον, ἀέρος στοιχεῖον. τρίτον
δὲ τὸ εἰκοσάεδρον βασίων μὲν εἴκοσι, γωνιῶν
δὲ δώδεκα, ὕδατος στοιχεῖον πολυμερέστερον
καὶ βαρύτερον. ταῦτα δ' ὦν ἀπὸ ταὐτῶ στοιχείω
συγκείμενα εἰς ἄλληλα τρέπεται. τὸ δὲ δωδεκάεδρον εἰκόνα τῶ παντὸς ἐστάσατο, ἔγγιστα
σφαίρᾳ ἐόν.

36. Πῦρ μὲν ὦν διὰ τὰν λεπτομέρειαν διὰ   98e
πάντων ἧκεν, ἀήρ τε διὰ τῶν ἄλλων ἔξω πυρός,
ὕδωρ δὲ διὰ τᾶς γᾶς. 37. ἅπαντα δ' ὦν πλήρη
ἐντί, οὐδὲν κενεὸν ἀπολείποντα. 38. συνάγεται
δὲ τᾷ περιφορᾷ τῶ παντός, καὶ ἠρεισμένα
τρίβεται μὲν ἀμοιβαδόν, ἀδιάλειπτον δὲ
ἀλλοίωσιν ποτὶ γενέσιας καὶ φθορὰς ἀποδίδωτι.

Because of this, earth is the heaviest and
most difficult of the elemental bodies to
move, an elemental body which does not
change into other elemental bodies because
it has nothing in common with the other
type of triangle. Only earth has its own
basic element, the isosceles triangle.

35. The other type of triangle is the
basic element of the other elemental bodies,
fire, air, and water.[16] When six of these
scalene triangles are put together, an
equilateral triangle is formed.[17] From
this a pyramid is put together, which has
four faces and as many corners, a most
mobile and finely divided form for fire.
After this comes the octahedron which has
eight faces and six corners and which is
the basic element for air. Third is the
icosahedron which has twenty faces and
twelve corners and which, being the more
highly divided and heavier, is the basic
element of water. Because all of these
are constructed from the same basic element,
they change into one another. He also
established the dodecahedron as the image
of the universe, since it is most like the
sphere.[18]

36. Therefore fire, because of its
fineness penetrates all of the others,
while air penetrates all but fire, and water
penetrates earth. 37. All things, therefore, are full, leaving nothing empty.
38. All things are held together by the
circular motion of the universe; and,
being pressed closely together, part rubs
against part and brings about continual
change through generation and dissolution.[19]

39. Τούτοις δὲ ποτιχρεόμενος ὁ θεὸς τόνδε
τὸν κόσμον κατεσκεύαξεν, ἁπτὸν μὲν διὰ τὰν   99a
γᾶν, ὁρατὸν δὲ διὰ τὸ πῦρ, ἅπερ δύο ἄκρα ἐντί.
δι' ἀέρος δὲ καὶ ὕδατος συνεδήσατο δεσμῷ
κρατίστῳ, ἀναλογίᾳ, ἃ καὶ αὐτὰν καὶ τὰ δι'
αὐτᾶς κρατεόμενα συνέχεν δύναται. 40. εἰ
μὲν ὦν ἐπίπεδον εἴη τὸ συνδεόμενον, μία
μεσότας ἱκανά ἐστιν· εἰ δέ κα στερεόν, δύο
χρήσει. δυσὶ δὴ μέσοις δύο ἄκρα ξυναρμόξατο,
ὅκως εἴη ὡς πῦρ ποτ' ἀέρα ἀὴρ ποτὶ ὕδωρ,
ὡς δέ κ' ἀὴρ ποτὶ ὕδωρ καὶ ὕδωρ ποτὶ γᾶν,   99b
καὶ κατ' ἐναλλαγάν, ὡς πῦρ ποτὶ ὕδωρ ἀὴρ
ποτὶ γᾶν, καὶ ἀνάπαλιν, ὡς γᾶ ποτὶ ὕδωρ
ὕδωρ ποτ' ἀέρα καὶ ἀὴρ ποτὶ πῦρ, καὶ κατ'
ἐναλλαγάν, ὡς γᾶ ποτ' ἀέρα ὕδωρ ποτὶ πῦρ.
41. καὶ ἐπεὶ δυνάμει ἴσα ἐντὶ πάντα, τοὶ
λόγοι αὐτῶν ἐν ἰσονομίᾳ ἐντί. εἷς μὲν ὦν
ὅδε ὁ κόσμος δαιμονίῳ δεσμῷ τῷ ἀνὰ λόγον
ἐστίν.
42. Ἕκαστον δὲ τῶν τετόρων σωμάτων
πολλὰ εἴδεα ἔχει. πῦρ μὲν φλόγα καὶ φῶς καὶ
αὐγάν, διὰ τὰν ἀνισότατα τῶν ἐν ἑκάστῳ αὐτῶν
τριγώνων· καττὰυτὰ δὲ καὶ ἀὴρ τὸ μὲν καθαρὸν   99c
καὶ αὖον, τὸ δὲ νοτερὸν καὶ ὁμιχλῶδες· ὕδωρ
δὲ τὸ μὲν ῥέον, τὸ δὲ παKτόν, ὁκόσον χιών τε
καὶ πάχνα χάλαζά τε καὶ κρύσταλλος· ὑγρῶν τε
τὸ μὲν ῥυτόν, ὡς μέλι, ἔλαιον, τὸ δὲ παKτόν,
ὡς πίσσα, κηρός· παKτῶ δὲ εἴδεα τὸ μὲν χυτὸν
χρυσός, ἄργυρος, χαλκός, κασσίτερος, μόλυβδος,
σταγών, τὸ δὲ θραυστὸν θεῖον, ἄσφαλτον,

39. By making use of these elemental
bodies, the god constructed this world,
which is tangible because of earth and
visible because of fire, the two basic
elements at the extremes. Through the use
of air and water, he bound it together
with a most powerful bond, proportion, which
is able to hold together both itself and
what is under its control. 40. Now, if
what is bound together were two-dimensional,
then one mean would be enough; but, since
it is three-dimensional, two will be needed.
He joined together the two extremes by two
means, so that, as fire is to air, air is to
water; and as air is to water, so water is
to earth; conversely, as fire is to water,
so air is to earth. The reverse is also
true; as earth is to water, so water is to
air and air to fire; conversely, as earth
is to air, so water is to fire.[20] 41. And
since all are equal in power, their ratios
are in equilibrium. This world, then, is
one because of a divine bond, proportion.

42. Each of the four elemental bodies
has many forms. Fire has the flame, light,
and "glowing," because of the inequality of
the triangles in each of them. In the same
way, air can be either clear and dry or
damp and foggy. Water can be either flowing
or frozen like snow and frost or hail and ice.
Some moist things are liquid, such as honey
and oil, while others are solid, such as
pitch and wax. Concerning types of solids,
some are fusible, such as gold, silver,
bronze, tin, lead, and copper, while others
are brittle, such as brimstone, asphalt,

νίτρον, ἅλες, στυπτηρία, λίθοι τοὶ ὁμογενέες.   99d
43. Μετὰ δὲ τὰν τῶ κόσμω σύστασιν ζώων
θνατῶν γένναοιν ἐμαχανάσατο, ἵν' ᾖ τέλεος
ποτὶ τὰν εἰκόνα παντελῶς ἀπειργασμένος.
44. τὰν μὲν ὦν ἀνθρωπίναν ψυχὰν ἐκ τῶν αὐτῶν
λόγων καὶ δυνάμίων συγκερασάμενος καὶ μερίξας
διένειμε τᾷ φύσει τᾷ ἀλλοιωτικᾷ παραδούς·
διαδεξαμένα δ' αὐτὸν ἐν τῷ ⟨γεννᾶν⟩ ἀπέργαζεν   99e
θνατά τε καὶ ἐφαμέρια ζῷα· 45. ὦν τὰς ψυχὰς
ἐπιρρύτως ἐνάγαγε τὰς μὲν ἀπὸ σελάνας, τὰς δ'
ἀπ' ἀλίω, τὰς δὲ ἀπὸ τῶν ἄλλων τῶν πλαζομένων ἐν τᾷ τῶ ἑτέρω μοίρᾳ, ἔξω μιᾶς τᾶς τοῦ
αὐτοῦ δυνάμιος, ἃν ἐν τῷ λογικῷ μέρει ἔμιξεν,
εἰκόνα σοφίας τοῖς εὐμοιρατοῦσι. 46. τᾶς
μὲν γὰρ ἀνθρωπίνας ψυχᾶς τὸ μὲν λογικόν ἐστι
καὶ νοερόν, τὸ δ' ἄλογον καὶ ἄφρον· τοῦ δὲ
λογικοῦ τὸ μὲν κρέσσον ἐκ τᾶς ταὐτοῦ φύσιος,
τὸ δὲ χέρηον ἐκ τᾶς τοῦ ἑτέρου. ἑκάτερον δὲ
περὶ τὰν κεφαλὰν ἵδρυται, ὡς τἄλλα μέρεα τᾶς
ψυχᾶς καὶ τῶ σώματος ὑπηρετεῖν τούτῳ, καθάπερ   100a
ὑπάτῳ τῶ σκάνεος ἅπαντος. τῶ δ' ἀλόγω μέρεος
τὸ μὲν θυμοειδὲς περὶ τὰν καρδίαν, τὸ δ'
ἐπιθυματικὸν περὶ τὸ ἧπαρ. 47. τοῦ δὲ
σώματος ἀρχὰν μὲν καὶ ῥίζαν μυελοῦ εἶμεν
ἐγκέφαλον, ἐν ᾧ ἁ ἁγεμονία.

---

8 ἀπέργαζεν N Baltes   ἀπεργάζετο Thesleff
et Marg   ἀπεργάζειν BE.

natron, salt, alum, and similar kinds of rocks.[21]   99d

43. After the establishment of the world, he began to plan the generation of mortal living beings, so that the world would be made complete in every way in relationship to the image.[22] 44. Then, when he had mixed and divided the human soul from the same proportions and powers,[23] he apportioned and handed it over to changeable nature. And nature, succeeding   99e
him in the process of generation, produced mortal and ephemeral living beings.
45. Then it (he?) introduced souls through infusion, some from the moon, others from the sun, still others from the other creatures formed in the portion of the Different, with the single exception of the power of the Same, which it (he?) mixed into the reasonable part as an image of wisdom for those whose portion is blessed.[24] 46. With regard to human souls, one part is reasonable and intelligent, but the other part is without reason and foolish. Of the reasonable, the superior element is of the nature of the Same, and the inferior element is of the nature of the Different. Both, however, are situated in the head so that the other parts of the soul and the body may serve it as one who is supreme over the whole body.[25]   100a
Of the non-rational part, the irascible element is located around the heart and the appetitive element around the liver.
47. The principal part of the body and the root of the spinal cord is the brain, in which is found the ruling power. And what

ἀπὸ δὲ τούτου οἷον ἀπόχυμα ῥεῖν διὰ τῶν
νωτίων σφονδύλων τὸ λοιπόν, ἐξ οὗ εἰς
σπέρμα καὶ γόνον μερίζεσθαι. ὀστέα δὲ μυελῶ
περιφράγματα. τουτέων δὲ σκέπαν εἶμεν τὰν   100b
σάρκα καὶ προκάλυμμα. συνδέσμοις δὲ ποττὰν
κίνασιν τοῖς νεύροις συνᾶψε τὰ ἄρθρα. τῶν δ᾽
ἐντοσθιδίων τὰ μὲν τροφᾶς χάριν, τὰ δὲ
σωτηρίας.

48. Κινασίων δὲ τῶν ἀπὸ τῶν ἐκτὸς τὰς
μὲν ἀναδιδομένας ἐς τὸν φρονέοντα τόπον
αἰσθήσιας εἶμεν· τὰς δ᾽ ὑπ᾽ ἀντίλαψιν μὴ
πιπτοίσας ἀνεπαισθήτως, ἢ τῷ τὰ πάσχοντα   100c
σώματα γεοειδέστερα εἶμεν, ἢ τῷ τὰς κινάσιας
ἀμενηνοτέρας γίγνεσθαι. 49. ὁκόσαι μὲν ὦν
ἐξίσταντι τὰν φύσιν, ἀλγειναί ἐντι· ὁκόσαι
δὲ ἀποκαθίσταντι ἐς αὐτάν, ἁδοναὶ ὀνυμαίνονται.

50. Τᾶν δ᾽ αἰσθησίων τὰν μὲν ὄψιν ἁμὶν
τὸν θεὸν ἀνάψαι εἰς θέαν τῶν ὠρανίων καὶ
ἐπιστάμας ἀνάλαψιν. 51. τὰν δ᾽ ἀκοὰν λόγων
καὶ μελῶν ἀντιλαπτικὰν ἔφυσεν· ἇς στερισκό-
μενος ἐκ γενέσιος ὁ ἄνθρωπος οὐδὲ λόγον ἔτι
προέσθαι δυνασεῖται. διὸ καὶ συγγενεστάταν   100d
τῷ λόγῳ ταύταν τὰν αἴσθασίν φαντι εἶμεν.
52. ὁκόσα δὲ πάθεα τῶν σωμάτων ὀνυμαίνεται,
ποτὶ τὰν ἁφὰν κλῄζεται, τὰ δὲ ῥοπᾷ ποτὶ τὰν
χώραν. ἃ μὲν γὰρ ἁφὰ κρίνει τὰς ζωτικὰς
δυνάμιας, θερμότατα, ψυχρότατα, ξηρότατα,
ὑγρότατα, λειότατα, τραχύτατα, εἴκοντα,

is left over from the brain, like something
poured out, flows through the spinal column
and is divided into semen and seed.[26] The
bones are enclosures for the marrow, and   100b
the flesh is their covering and protective
layer. He bound the joints together with
tendons for the purpose of motion. Of the
internal organs, some are for nourishment
and others for preservation.

    48. Of motions that come from external
objects, those that are delivered to where
the understanding is located are sense per-
ceptions; those that do not fall within
apprehension remain imperceptible because
either the bodies affected are too earth-   100c
filled or the motions are too weak. 49.
Whatever motions, then, do violence to
something's nature are pains, and those
that restore it to its proper nature are
called pleasures.

    50. As for types of sense perception,
the god has fastened sight to us for the
contemplation of the heavens and the acqui-
sition of knowledge. 51. He produced hearing for the apprehension of words and
melodies; one who is deaf from birth will
later be unable to utter words. Therefore,   100d
this sense is said to be the one most closely
related to speech. 52. What we call the
qualities of elemental bodies are so named
because of their relationship to touch,
although also because of their inclination
toward a certain place. For touch distinguishes the properties of living beings,
warmth or coldness, dryness or moistness,
smoothness or roughness, flexibility or

ἀντίτυπα, μαλακά, σκληρά. 53. βαρὺ δὲ καὶ
κοῦφον ἁφᾷ μὲν προκρίνει, λόγος δ᾽ ὁρίζει
τᾷ ποτὶ τὸ μέσον καὶ ἀπὸ τῶ μέσω νεύσει.
54. κάτω δὲ καὶ μέσον ταὐτόν φαντι· τὸ γὰρ   100e
κέντρον τᾶς σφαίρας τοῦτό ἐντι τὸ κάτω, τὸ
δ᾽ ὑπὲρ τούτω ἄχρι τᾶς περιφερείας ἄνω.
55. τὸ μὲν ὦν θερμὸν λεπτομερές τε καὶ δια-
στατικὸν τῶν σωμάτων δοκεῖ εἶμεν, τὸ δὲ
ψυχρὸν παχυμερέστερόν τε τῶν πόρων καὶ
συμπιλωτικόν ἐστι. 56. τὰ δὲ περὶ τὰν
γεῦσιν ἔοικε τᾷ ἁφᾷ· συγκρίσει γὰρ καὶ
διακρίσει, ἔτι δὲ τᾷ ἐς τὼς πόρως διαδύσει
καὶ τοῖς σχημάτεσσιν ἢ στρυφνὰ ἢ λεῖα· ἀπο-
τάκοντα μὲν καὶ ῥύπτοντα τὰν γλῶτταν στρυφνὰ
φαίνεται, μετριάζοντα δὲ τᾷ ῥύψει ἀλμυρά,
ἐκπυροῦντα δὲ καὶ διαιρέοντα τὰν σάρκα δριμέα,   101a
τὰ δ᾽ ἐναντία λεῖά τε καὶ γλυκέα, και χυλῶται.
57. ὀσμᾶς δὲ εἴδεα μὲν οὐ διώρισται· διὰ γὰρ
στενῶν πόρων διαθεῖσθαι στερροτέρων ὄντων ἢ
ὡς συνάγεσθαι καὶ διίστασθαι σάψεσι δὲ καὶ
πέψεσι γᾶς τε καὶ γεοειδέων εὐώδεά τε καὶ
δυσώδεα εἶμεν.
58. Φωνὰ δ᾽ ἐστὶ μὲν πλᾶξις ἐν ἀέρι
διικνουμένα ποτὶ τὰν ψυχὰν δι᾽ ὤτων· ὦν τοὶ
πόροι διήκοντι ἄχρις ἥπατος χωρέοντες, ἔν τ᾽
αὐτοῖς πνεῦμα, οὗ ἁ κίνασις ἀκοά ἐστι. φωνᾶς
δὲ καὶ ἀκουᾶς ἁ μὲν ταχεῖα   101b

rigidity, softness or hardness. 53. Touch
makes a preliminary distinction between heavy
and light, but reason defines it by the
object's inclination either toward the middle
or away from the middle.[27] 54. "Below" and   100e
"middle" are said to be the same. The
center of a sphere is its "below," and whatever is beyond it as far as the circumference is "above." 55. Heat, then, appears to
consist of small particles and divides
bodies; cold, however, is made up of particles
too coarse for movement through the passages
and is apt to compress them. 56. What happens with taste is similar to what happens
with touch. For things are perceived as
either astringent or smooth because of con-
traction and dilation, as well as because
of their movement through the passages and
their shapes. Things that melt and cleanse
the tongue are perceived as astringent, and
things more moderate in terms of cleansing
are perceived as salty. Things that burn
and divide the flesh are perceived as pun-   101a
gent, and the opposite things as smooth
and sweet and juicy. 57. Types of odors
are not distinguished because they are forced
through narrow passages that are too rigid to
either shrink or expand. Pleasant odors and
foul odors are caused by the decay and
ripening of earth and of things like earth.

58. Sound is a percussion in the air
that penetrates to the soul through the ears.
The passages from the ears continue on and
extent as far as the liver, and in those
passages is *pneuma*, whose motion constitutes
hearing.[28] As for sound and hearing, a rapid   101b

ὀξεῖα, ἃ δὲ βραδεῖα βαρεῖα, μέσα δ' ἃ συμ-
μετροτάτα. καὶ ἃ μὲν πολλὰ καὶ κεχυμένα
μεγάλα, ἃ δὲ ὀλίγα καὶ συναγμένα μικρά.
ἃ δὲ τεταγμένα ποτὶ λόγως μωσικῶς ἐμμελής,
ἃ δὲ ἄτακτός τε καὶ ἄλογος ἐκμελής τε καὶ
ἀνάρμοστος. 59. τέταρτον δὲ γένος αἰσθητῶν
πολυειδέστατον καὶ ποικιλώτατον, ὁρατὰ δὲ
λέγεται· ἐν ᾧ χρώματά τε παντοῖα καὶ κεχρωσ-
μένα μυρία, πρᾶτα δὲ τέτορα, λευκόν, μέλαν,   101c
λαμπρόν, φοινικοῦν· τἆλλα γὰρ ἐκ κιρναμένων
τούτων γεννᾶται. τὸ μὲν ὦν λευκὸν διακρίνει
τὰν ὄψιν, τὸ δὲ μέλαν συγκρίνει, ὅκως περ τὸ
μὲν θερμὸν διαχῆν τὰν ἁφάν, τὸ δὲ ψυχρὸν
συνάγεν δύναται, καὶ τὸ μὲν στρυφνὸν συνάγεν
τὰν γεῦσιν, τὸ δὲ δριμὺ διαιρῆν πέφυκε.

60. Τρέφεσθαι δὲ τὸ σκᾶνος τῶν ἐναερίων
ζώων καὶ συνέχεσθαι τᾶς μὲν τροφᾶς διαδιδο-
μένας διὰ τῶν φλεβῶν εἰς ὅλον τὸν ὄγκον κατ'
ἐπιρροάν, οἷον δι' ὀχετῶν ἀγομένας, καὶ ἀρδο-   101d
μένας ὑπὸ τῶ πνεύματος, ὃ διαχεῖ αὐτὰν ἐπὶ
τὰ πέρατα φέρον.

61. Ἁ δ' ἀνάπνοια γίνεται μηδενὸς μὲν
κενοῦ ἐν τᾷ φύσει ἐόντος, ἐπιρρέοντος δὲ καὶ
ἑλκομένω τῶ ἀέρος ἀντὶ τῶ ἀπορρέοντος διὰ
τῶν ἀοράτων στομίων, δι' ὦν καὶ ἁ νοτὶς ἐπι-
φαίνεται· τινὸς δὲ καὶ ὑπὸ τᾶς φυσικᾶς
θερμότατος ἀπαναλωμένω. 62. ἀνάγκα ὦν

motion produces a high-pitched sound, a
slow motion produces a low-pitched one,
and the most balanced motion, a medium
pitch. A large and extended motion is
loud, while a small and restricted one is
soft. Those arranged according to musical
intervals are melodic, but those not arranged
in proper intervals are unmelodic and un-
harmonic. 59. The fourth kind of sensible
object is the most diverse and varied and
is called the visible. In this type there
are all kinds of colors and innumerable
colored objects, but there are four primary
colors: white, black, bright, and red.[29]
For the other colors are generated by
mixing the primary colors. White, then,
dilates the eye but black contracts it,
just as heat relaxes but cold contracts
touch, and just as the astringent contracts
but the pungent divides taste.[30]

60. The body of living beings that
dwell in air is nourished and held together
because nourishment is distributed in
streams through the veins to the whole
bodily mass, as if brought through water   101d
channels, and is irrigated by *pneuma*,
which disperses it and brings it to the
body's outermost limits.

61. Since there is no void in nature,
breathing occurs when air flows and is
drawn in as other air flows out through
invisible openings, openings through which
perspiration also appears. Part of the air,
however, is used up by the natural heat
involved. 62. It is necessary, therefore,

ἀντικαταχθῆμεν τὸ ἴσον τῷ ἀναλωθέντι· εἰ δὲ
μή, κενώσιας εἶμεν, ὅπερ ἀμάχανον· οὐδὲ γὰρ
ἔτι εἴη κα σύρροον καὶ ἓν τὸ ζῶον, διαιρωμένω
τῶ σκάνεος ὑπὸ τῶ κενῶ. 63. ἁ δ' ὁμοία   101e
ὀργανοποιία γίγνεται καὶ ἐπὶ τῶν ἀψύχων
221   καττὰν τᾶς ἀναπνοᾶς ἀναλογίαν· ἁ γὰρ σικύα
καὶ τὸ ἤλεκτρον εἰκόνες ἀναπνοᾶς ἐντι.
64. ῥεῖ γὰρ διὰ τῶ σώματος ἔξω θύραζε τὸ   102a
πνεῦμα, ἀντεπεισάγεται δὲ διὰ τᾶς ἀναπνοᾶς
τῷ τε στόματι καὶ ταῖς ῥισίν, εἶτα πάλιν
οἷον Εὔριπος ὃ μὲν ἀντεπιφέρεται εἰς τὸ σῶμα,
ὃ δὲ ἀνατείνεται καττὰς ἐκροάς. 65. ἁ δὲ
σικύα ἀπαναλωθέντος ὑπὸ τῶ πυρὸς τῶ ἀέρος
ἐφέλκεται τὸ ὑγρόν, τὸ δ' ἤλεκτρον ἐκκριθέν-
τος τῶ πνεύματος ἀναλαμβάνει τὸ ὁμόριον σῶμα.

66. Τροφὰ δὲ πᾶσα ἀπὸ ῥίζας μὲν τᾶς
καρδίας, παγᾶς δὲ τᾶς κοιλίας ἐπάγεται τῷ
σώματι· καὶ εἴ κα πλείω τᾶς ἀπορρεοίσας   102b
ἐπάρδοιτο, αὔξα λέγεται, εἴ κα δὲ μείω,
φθίσις· ἁ δ' ἀκμὰ μεθόριόν τε τουτέων ἐστὶ
καὶ ἐν ἰσότατι ἀπορροᾶς καὶ ἐπιρροᾶς νοέεται.
67. λυομένων δὲ τῶν ἁρμῶν τᾶς συστάσιος, αἳ
κα μηκέτι δίοδος ᾖ πνεύματι ἢ τροφᾷ διαδιδῶ-
ται, θνάσκει τὸ ζῶον.

68. Πολλαὶ δὲ κᾶρες ζωᾶς καὶ θανάτου
αἰτίαι. ἓν δ' ὧν γένος νόσος ὀνυμαίνεται.
νόσων δ' ἀρχαὶ μὲν αἱ τᾶν πρατᾶν δυναμίων
ἀσυμμετρίαι, εἴκα πλεονάζοιεν ἢ ἐλλείποιεν
ταὶ ἁπλαῖ δυνάμιες, θερμότας ἢ ψυχρότας ἢ   102c
ὑγρότας ἢ ξηρότας.

---

3 κα σύρροον vulg. (cf. infra. p. 76.22)
κα σύνσοον N   κασσύρροον BWV.

11 ὃ μὲν ἀντεπιφέρεται coni. Baltes
ὃ μὲν omm. codd.

12 ὃ δὲ ἀνατείνεται B Baltes
τὸ δὲ ἀνατείνεται NWV.

15 ὁμόριον coni. Baltes   ὅμοιον NBWV.

that a like amount take the place of what
was used up. Otherwise empty spaces result,
which is an impossibility. For the living
being would no longer flow together smoothly
and be a unity, since the body would be
divided by the void. 63. The same kind of   101e
functional structure exists also in soulless
beings, based on the analogy of breathing.
64. *Pneuma* flows outward through the body   102a
but is brought back in again by breathing
through the mouth and nostrils; like the
(ebb and flow of the) Euripus strait, some
pneuma rushes back into the body while
other *pneuma* moves upward toward the outlets.[31] 65. When air is used up by fire,
the cupping-glass draws out the moisture;
and amber, when *pneuma* has been expelled,
attracts the body nearby.[32]

66. All nourishment is supplied to the
body from the heart as the root and from
the abdomen as the source. Whenever the   102b
body is irrigated more than it is drained,
it is known as growth; whenever less, it
is known as decay. The zenith is the
borderline between these two and is seen in
the equilibrium of outflow and intake.
67. Once the fastenings of this structure
are loosened, whenever there is no longer
passage for *pneuma* or nourishment is no
longer distributed, the living being dies.

68. There are many threats to life and
causes of death. One type is called disease.
The basic cause of disease is lack of proportion among the primary properties, whenever the basic properties, such as heat or
cold, moisture or dryness, are excessive   102c

69. μετὰ δὲ ταύτας αἱ τῶ αἵματος τροπαὶ καὶ
ἀλλοιώσιες ἐκ διαφθορᾶς καὶ αἱ τᾶς σαρκὸς
τακομένας κακώσιες, αἱ καττὰς μεταβολὰς ἐπὶ
τὸ ὀξὺ ἢ ἁλμυρὸν ἢ δριμὺ τροπαὶ αἵματος ἢ
σαρκὸς τακεδόνες γένοιντο. χολᾶς γὰρ αἱ
γενέσιες καὶ φλέγματος ἐνθένδε χυμοί τε
νοσώδεες καὶ ὑγρῶν σάψιες, ἀμαυραὶ μὲν αἱ
μὴ ἐν βάθει, χαλεπαὶ δ' ὧν ἀρχαὶ γεννῶνται
ἐξ ὀστέων, ἀνίατοι δὲ ἐκ μυελοῦ ἐξαπτόμεναι.   102d
70. τελευταία δὲ νόσων αἰτία ἐντὶ πνεῦμα,
χολά, φλέγμα αὐξόμενα καὶ ῥέοντα ἐς χώρας
ἀλλοτρίας ἢ τόπως ἐπικαιρίως. τόκα γὰρ ἀντι-
καταλαμβάνοντα τὰν τῶν καρρόνων χώραν καὶ
ἀπελάσαντα τὰ συγγενέα ἱδρύεται κακοῦντα τὰ
σώματα καὶ ἐς αὕταυτα ἀναλύοντα. καὶ σώματος
μὲν πάθεα τάδε τε καὶ ἐκ τῶνδε.

71. Ψυχᾶς δὲ νόσοι ἐντὶ πολλαί, ἄλλαι δ'
ἄλλων δυναμίων ἐντί, αἰσθητικᾶς μὲν δυσαισθησία,
μναμονικᾶς δὲ λάθα, ὁρματικᾶς δὲ ἀνορεξία τε   102e
καὶ προπέτεια, παθητικᾶς δὲ ἄγρια πάθεά τε
καὶ λύσσαι οἰστρώδεες, λογικᾶς δὲ ἀμαθία καὶ
ἐκφροσύνα. 72. ἀρχαὶ δὲ κακίας ἀδοναὶ καὶ
λῦπαι ἐπιθυμίαι τε καὶ φόβοι, ἐξαμμέναι μὲν
ἐκ σώματος, ἀνακεκραμέναι δὲ τᾷ ψυχᾷ· καὶ
ἐξαγγελλόμεναι ὀνόμασι ποικίλοις· ἔρωτες γὰρ
καὶ πόθοι ἵμεροί τε ἔκλυτοι

---

9 ἀνίατοι coni. Baltes   ἀνιαραὶ NWV   ἔνιαι B.

or deficient. 69. After these come the
changes and alterations of the blood
caused by deterioration, and the injuries
to the flesh as it decays, whenever changes
in the blood or decay of the flesh result
from shifts to the astringent, the salty,
or the pungent. For from this comes the
formation of bile and phlegm as well as
unwholesome humors, and the putrifaction
of fluids. Their effects are mild when they
are not deeply imbedded, but serious when
the causes are situated in the bones, and
incurable when situated in the marrow. 102d
70. The final cause of disease is *pneuma*,
bile, or phlegm increasing and flowing into
abnormal areas or vital places. For, then,
they establish themselves by taking the
place of better substances and driving
kindred substances out,[33] and so harm
bodies and dissolve them into themselves.
These, then, are the ills of the body and
their causes.

71. Diseases of the soul are many,
and different faculties of the soul are
affected by different diseases: for the
faculty of sense perception, the disease
is dull sense perception; for the faculty
of memory, forgetfulness; for the appetitive 102e
faculty, wild passions as well as frantic
rages; and for the faculty of reason, stu-
pidity and madness. 72. The causes of vice
are pleasures and pains, desires and fears,
all of which depend on the body but which
are also mixed with the soul. They are
known by a variety of names. For there are
loves and longings, unbridled passions,

ὀργαί τε σύντονοι καὶ θυμοὶ βαρεῖς ἐπιθυμίαι
τε ποικίλαι καὶ ἀδοναὶ ἄμετροί ἐντι.
73. ἁπλῶς δὲ εἰπεῖν· τὸ πῶς ἔχεν ποτὶ τὰ
πάθεα ἀρχά τε καὶ πέρας ἀρετᾶς καὶ κακίας
ἐστί· τὸ γὰρ πλεονάζεν ἐν ταύταις ἢ κάρρον
αὐτᾶν εἶμεν εὖ ἢ κακῶς ἁμὲ διατίθητι.
74. ποτὶ δὲ ταύτας τὰς ὁρμὰς μεγάλα μὲν
συνεργῆν δύνανται αἱ τῶν σωμάτων κράσιες,
ὀξεῖαι ἢ θερμαὶ ἢ ἄλλοτ᾽ ἀλλοῖαι γιγνόμεναι,
ἔς τε μελαγχολίας καὶ λαγνείας καὶ λαβρότατας
ἄγοισαι ἁμέ. 75. καὶ ῥευματιζόμενά τινα
μέρεα ὀδαξασμὼς ποιέντι καὶ μορφὰς φλεγμαι-   103b
νόντων σωμάτων μᾶλλον ἢ ὑγιαινόντων, δι᾽ ὧν
δυσθυμίαι καὶ λᾶθαι παραφροσύναι τε καὶ
πτοῖαι ἀπεργάζονται. 76. ἱκανὰ δὲ τὰ ἔθεα,
ἐν οἷς ἂν ἐντραφῶσι κατὰ πόλιν ἢ οἶκον, καὶ ἁ
καθ᾽ ἁμέραν δίαιτα θρύπτοισα τὰν ψυχὰν ἢ
ῥωννύοισα ποτ᾽ ἀλκάν· ταὶ γὰρ θυραυλίαι καὶ
ἁπλαῖ τροφαὶ τά τε γυμνάσια καὶ τὰ ἤθεα τῶν
συνόντων τὰ μέγιστα δύνανται ποτὶ ἀρετὰν καὶ
ποτὶ κακίαν. 77. κατὰ ταῦτα κακίας μὲν αἰτία
ἐκ τῶν γενετόρων καὶ στοιχείων ἐπάγεται   103c
μᾶλλον ἢ ἐξ ἁμέων· ὃ τι μὴ ἀργία ἐστίν, ἀφι-
σταμένων ἡμῶν τῶν ποθακόντων ἔργων.
78. Ποτὶ δὲ τὸ εὖ ἔχεν τὸ ζῷον δεῖ τὸ
σῶμα ἔχεν τὰς ὑπ᾽ αὐτῷ ἀρετάς, ὑγείαν τε καὶ
εὐαισθησίαν ἰσχύν τε καὶ κάλλος. ἀρχὰ δὲ
κάλλους συμμετρία ποτί τ᾽ αὐτῶ τὰ μέρεα καὶ
ποτὶ τὰν ψυχάν· ἁ γὰρ φύσις οἷον ὄργανον
ἁρμόξατο τὸ σκᾶνος, ὑπάκουόν τε εἶμεν καὶ

---

5 ταύταις codd.   τούτοις coni. Marg
6 αὐτᾶν M² Baltes   αὐτον B   αὐτᾶς NWV
αὐτῶν coni. Marg.

Text and Translation

acts of vehement anger and violent temper, assorted desires and immoderate pleasures. 73. Put simply, the way in which a person deals with the passions is the beginning and end of virtue and vice. For whether we are excessive in them or control them establishes us in a state of being well or ill.[34] 74. The bodily mixtures, whether astringent, sharp, or warm, or at other times changing, can greatly affect these impulses and so lead us into forms of melancholy, lust, and violence.[35] 75. And some parts, when they suffer from a flux, produce itches and types of inflamed rather than healthy bodies, from which despair and forgetfulness, delirium and fright result. 76. The habits in which a person is raised either in the city or at home are important, and one's daily lifestyle either enfeebles the soul or makes it courageous. For this reason camping out and simple foods, exercise and the habits of one's companions are extremely important, both for virtue and for vice. 77. Accordingly, the cause of vice comes rather from our parents and from our own basic elements than from ourselves, granted that there is no laziness and that we do not shrink from our proper duties.

78. In order for a living being to be well, its body must have the virtues proper to it, health and keen sense perception, strength and beauty. The basis of (bodily) beauty is symmetry both in relationship to its parts and in relationship to the soul. For nature has constructed this "tent" like an instrument that is to be obedient to and

ἐναρμόνιον ταῖς τῶν βίων ὑποθέσεσι. 79. δεῖ 103d
δὲ καὶ τὰν ψυχὰν ῥυθμίζεσθαι ποτὶ τὰς ἀνα-
λόγως ἀρετάς, ποτὶ μὲν σωφροσύναν οἷον ποτὶ
ὑγείαν τὸ σῶμα, ποτὶ δὲ φρόνασιν οἷον ποτὶ
εὐαισθασίαν, ποτὶ δὲ ἀνδρειότατα οἷον ποτὶ
ῥώμαν καὶ ἰσχύν, ποτὶ δὲ δικαιοσύναν οἷον
ποτὶ κάλλος τὸ σῶμα. 80. τουτέων δὲ ἀρχαὶ
μὲν ἐκ φύσιος, μέσα δὲ καὶ πέρατα ἐξ ἐπιμελείας,
σώματος μὲν διὰ γυμναστικᾶς καὶ ἰατρικᾶς,
ψυχᾶς δὲ διὰ παιδείας καὶ φιλοσοφίας. αὗται
γὰρ ταὶ δυνάμιες τρέφοισαί τε καὶ τονοῖσαι
καὶ τὰ σώματα καὶ τὰς ψυχὰς διὰ πόνων καὶ 103e
γυμνασίων καὶ διαίτας καὶ ὀρνύοντι ὅκα δεῖ,
ταὶ μὲν διὰ φαρμακειᾶν, ταὶ δὲ παιδευτικαὶ
τᾶν ψυχᾶν διὰ κολασίων καὶ ἐπιπλαξίων.
ῥωννύοντι δὲ καὶ διὰ προτροπᾶν ἐγείροισαι
τὰν ὁρμὰν καὶ ἐγκελευόμεναι τὰ ποτίφορα
ποττὰ ἔργα. 81. ἀληπτικὰ μὲν ὦν καὶ ἃ ταύτᾳ 104a
συγγενεστάτα ἰατρικά, σώματα ταχθεῖσαι θερα-
πεύεν, ἐς τὰν κρατίσταν ἁρμονίαν ἄγοισαι
τὰς δυνάμιας τό τε αἷμα καθαρὸν καὶ τὸ
πνεῦμα σύρροον ἀπεργάζονται, ἵν' εἰ καί τι
νοσῶδες ὑπογένοιτο, κράτος αὐτοῦ ἔχοιεν
ἐρρωμέναι ταὶ δυνάμιες αἵματος καὶ πνεύματος.
82. Μωσικὰ δὲ καὶ ἃ ταύτας ἁγεμὼν φιλοσοφία, ἐπὶ τᾷ τᾶς ψυχᾶς ἐπανορθώσει ταχθεῖσαι 104b
ὑπὸ θεῶν τε καὶ νόμων, ἐθίζοντι καὶ πείθοντι,
τὰ δὲ καὶ ποταναγκάζοντι, τὸ μὲν ἄλογον τῷ
λογισμῷ πείθεσθαι, τῶ δ' ἀλόγῳ θυμὸν μὲν

---

27 θεῶν AL WV   ἐθέων N (coni. Valckenaer)
28 ἄλογον N AL WV   λογικὸν coni. Marg.

in harmony with the purposes of various types of life.[36] 79. The soul also must be trained for the corresponding virtues: for temperance, as the body is trained for health; for prudence, as the body is trained for keen sense perception; for courage, as the body is trained for might and strength; and for justice, as the body is trained for beauty. 80. The beginnings of these virtues come from nature, but the middle and final stages come from diligent effort, through exercise and medicine for the body and <u>through education and philosophy for the soul</u>. These powers nourish and brace up both bodies and souls through various hardships, exercises, and regimens; and, when necessary, they even arouse them, some through purgatives, and others, which educate the soul, by means of punishments and rebukes. They strengthen them also, both inciting them by encouragement to make the attempt and urging them on to appropriate deeds. 81. Training therefore, and that most closely related to it, medicine, both ordered to the care of the body, make the blood clean and the *pneuma* flow smoothly together by bringing their properties into the most harmonious relationship, so that, if a sickly condition does arise, the now strengthened properties of the blood and *pneuma* would be able to overcome it.

82. Music and philosophy, <u>its guide</u>, which were established by the gods and the laws for the correction of the soul, accustom, persuade, and sometimes even coerce the non-rational part to obey reason, the irascible part of the non-rational soul to

103d

103e

104a

104b

πρᾶον εἶμεν, ἐπιθυμίαν δὲ ἐν ἀρεμήσει, ὡς μὴ
δίχα λόγω κινέεσθαι, μηδὲ μὰν ἀτρεμίζειν τῶ
νῶ ἐκκαλεομένω ἢ ποτὶ ἔργα ἢ ποτὶ ἀπολαύσιας.
οὗτος γάρ ἐστιν ὅρος σωφροσύνας εὐπείθειά τε
καὶ καρτερία. 83. καὶ σύνεσις καὶ ἁ πρεσβίστα
φιλοσοφία, ἀποκαθαράμεναι ψευδέας δόξας,
ἐνέθηκαν τὰν ἐπιστάμαν, ἀνακαλεσάμεναι τὸν   104c
νόον ἐκ μεγάλας τᾶς ἀγνοίας, χαλάσασαι ἐς ὄψιν
τῶν θείων, τοῖς ἐνδιατρίβεν σὺν αὐταρκείᾳ τε
ποττἀνθρώπεια καὶ σὺν εὐροίᾳ ἐπὶ τὸν σύμμετρον
βίω χρόνον εὐδαιμονεῖν ἐστιν. ὅτῳ μὲν δὴ ὁ
δαίμων μοίρας τάσδ' ἔλαχε, δι' ἀλαθεστάταν
δόξαν ἄγεται ἐπὶ τὸν εὐδαιμονέστατον βίον.

84. Εἰ δέ κά τις σκλαρὸς καὶ ἀπειθής,
τούτῳ δὲ ἐπέσθω κόλασις ἅ τ' ἐκ τῶν νόμων καὶ
ἁ ἐκ τῶν λόγων, σύντονα ἐπάγοισα δείματά τε
ὑπωράνια καὶ τὰ καθ' "Αιδεος, ὅθι κολάσιες
ἀπαραίτητοι ἀπόκεινται δυσδαίμοσι νερτέροις,   104d
καὶ τἆλλα ὅσα ἐπαινέω τὸν 'Ιωνικὸν ποιητὰν
ἐκπλαγέντας ποιεῦντα τὼς ἐναγέας. 85. ὡς γὰρ
τὰ σώματα νοσώδεσί ποκα ὑγιάζομες, αἴ κα μὴ
εἴκῃ τοῖς ὑγιεινοτάτοις, οὕτω τὰς ψυχὰς
ἀνείργομες ψευδέσι λόγοις, αἴ κα μὴ ἄγηται
ἀλαθέσι. 86. λέγοιντο δ' ἀναγκαίως καὶ
τιμωρίαι ξέναι, ὡς μετενδυομενᾶν τᾶν ψυχᾶν
τῶν μὲν δειλῶν ἐς γυναικεῖα σκάνεα μοθ'
ὕβριν ἐκδιδόμενα, τῶν δὲ μιαιφόνων ἐς θηρίων   104e
σώματα

be tame, and the appetitive part to remain
quiet when the mind summons it either to
action or to enjoyment.³⁷ For this is the
mark of temperance: docility and steadfast-
ness. 83. In addition, intelligence and
the highest philosophy, by purifying away
false opinions, have established knowledge,
calling the mind up from great ignorance   104c
and releasing it for the vision of divine
realities. To occupy oneself with these
for a suitable period of one's life, independent of human affairs and in a state of
well-being, is to find happiness. The one
to whom the daimon grants such a fate is led
by truest opinion to the happiest kind of
life.³⁸

implies wealth

To whom?

84. But if anyone is stubborn and dis-
obedient, let punishment according to both
the laws and the traditional stories come
upon him, punishment that brings with it
intense terrors both under the heavens and
in Hades, where inexorable punishments are
reserved for the unlucky dead, as well as   104d
those other terrors for which I praise the
Ionian poet who struck the cursed with
fear.³⁹ 85. For, just as we sometimes restore bodies to health by means of sicknesses,
if someone does not follow really healthy
ways of life, so too we restrain souls with
false stories, if someone will not be guided
by true ones. 86. Unusual punishments must
surely be noted, in that the souls of the
cowardly are clothed in female bodies which
are given over to lust; the souls of mur-   104e
derers are clothed in the bodies of beasts

Afterlife!

ποτὶ κόλασιν, λάγνων δ' ἐς ὄνων ἢ κάπρων
μορφάς, κούφων δὲ καὶ μετεώρων ἐς πτηνῶν
ἀεροπόρων, ἀργῶν δὲ καὶ ἀπράκτων ἀμαθῶν τε
καὶ ἀνοήτων ἐς τὰν τῶν ἐνύδρων ἰδέαν.
87. Ἅπαντα δὲ ταῦτα ἐν δευτέρᾳ περιόδῳ
ἁ Νέμεσις συνδιέκρινε σὺν δαίμοσι παλαμναίοις   105a
χθονίοις τε, τοῖς ἐπόπταις τῶν ἀνθρωπίνων,
88. οἷς ὁ πάντων ἀγεμὼν θεὸς ἐπέτρεψε διοίκησιν
κόσμω συμπεπληρωμένω ἐκ θεῶν τε καὶ ἀνθρώπων
τῶν τε ἄλλων ζώων, ὅσα δεδαμιούργηται ποτ'
εἰκόνα τὰν ἀρίσταν εἴδεος ἀγεννήτω καὶ αἰωνίω
καὶ νοατῶ.

for punishment; those of the debauchers are clothed in the shapes of asses and boars; those of the lightheaded and thoughtless are clothed in the bodies of birds that ply the air; and those of the idle and indolent, as well as those of the ignorant and foolish, are clothed in the shape of water creatures.[40]

87. Nemesis, together with the avenging and chthonic daimons, the overseers of human affairs, have determined all of these things for the second cycle (of birth). 88. The god, the ruler of all, turned over to them the direction of the world which is filled with gods and human beings as well as with other living beings. And all of these beings were fashioned in relationship to the best-possible image of the ungenerated, eternal, and intelligible form.[41]

105a

Efficiency is that of Gods!

NOTES TO THE TRANSLATION

[1] The term λόγος is consistently used in the TL in the sense of "proportion" or "ratio"; see TL 14, 15, 19, 41.

[2] By this point in the TL, it is clear that matter rather than Necessity is one of the two basic elements in the construction of the sensible world (the other being the idea). The use of matter as a metaphysical principle first appears in Aristotle (e.g. *Meta*. 7-8, 1028a-1045b) and is common in Middle Platonism (e.g. Plutarch, *De proc. an. in Tim.* 24, 1024C; *De Is. et Os.* 53, 372F).

[3] It is clear from TL 11 that this world is a limit only in the realm of sensible being. There also exists another realm, the intelligible, beyond that of the sensible.

[4] The καί is explanatory. In TL 3 the idea has been identified with the intelligible substance.

[5] For example: a : b :: b : c; the reverse: c : b :: b : a; the converse: b : a :: c : b. Justice (Δίκα) in this context may simply be the personification of the laws of the universe. But Justice is also the Pythagorean name for three and so probably is used here in connection with the relationships of these *three* terms among themselves. See Plutarch, *De Is. et Os.* 75, 381F-382A.

[6] The author of the TL seems to be saying that the world-soul is first of all a mixture of the idea (the indivisible pattern) and matter (the divisible substance). This interpretation of *Tim*. 35a is also found in Albinus (*Didaskalikos* 14, p. 169, 20ff.).

[7] This table is Marg's reconstruction of what originally stood in the text of the TL. See Marg, ed., *De natura mundi et animae*, 60-78, 124-31, and the present introduction, pp. 21-22.

[8] "Their generator" or "the generator of *these* things," meaning the world of sensible things. The latter translation would mean that the author of the TL distinguishes between the god's relationship to the ideal world (originator) and his relationship to the sensible world (generator).

[9] The point seems to be that, while we can see an individual sensible object with our eyes, we can come to understand the sensible world as a whole only through contemplation of the heavenly bodies, that is, through astronomy. Yet even here our view remains partial.

[10] The movement of the universe as a whole is from east to west and is identical with the motion of the Same.

[11] The popular use of the name "Lightbringer" for Venus implies that Venus is *only* the morning star and that the *only* morning star is Venus. The author of the TL, however, wants to show that both of these popular notions are wrong. Venus can also appear in the evening, and other stars can also precede the rising of the sun and so be "light-bringers."

[12] Its own proper motion is the motion of the Different.

[13] We define the "setting" or "rising" of a heavenly body by what we can see from a certain point on the surface of the earth in relationship to the horizon as seen from that particular point.

[14] "Isosceles triangle," literally a "half square," because it forms half of a square (TL 33).

[15] The construction of the square from *four* isosceles triangles rather than from *two* follows *Tim.* 55b. Why Plato did this has always been a puzzle. A similar puzzle is found in *Tim.* 54d-e (TL 35) where the equilateral triangle is constructed from *six* scalene triangles rather than from *two*. Cornford (*Plato's Cosmology*, 230-34) suggests that Plato thought each elemental body (earth, water, air, and fire) came in several sizes (*Tim.* 58c-d; TL 42). However, Plato explicitly describes only the construction of the intermediate size. In addition, Cornford (234-39) suggests that Plato's construction of these figures is similar to several proofs found in Euclid (13.12-13), proofs discovered earlier, perhaps by Theaetetus, and known by Plato.

[16] The other type of triangle is the scalene triangle.

[17] "Scalene triangle," literally a "half triangle," because it forms *half* of a bisected equilateral triangle (TL 33). See note 14.

[18] The reason the dodecahedron seems most like the sphere is given in *Phd.* 110b, where Socrates describes the earth, viewed from above, as looking like "one of those balls made from twelve pieces of leather." The ball was made from twelve pieces of leather, each in the shape of a regular pentagon. If the leather were stiff, it would form a regular dodecahedron; because it is flexible, it forms a ball. This explanation is found in Plutarch (*Quaest. Plat.* 5.1, 1003C-D).

[19] Although there is continual change, an overall equilibrium among the elemental bodies is always maintained. See TL 41.

[20]*basic proportion:* a : b :: b : c :: c : d
converse of basic proportion: a : c :: b : d
reverse of basic proportion: d : c :: c : b :: b : a
converse of the reverse: d : b :: c : a

[21]The subdivisions of fire and air are basically the same as those of *Tim.* 58c-d. However, the subdivisions of water and earth differ substantially from those of *Tim.* 58d-61c. The major differences are two: (1) the TL breaks "water" down into "water" and "moist things"; and (2) the TL treats fusible metals with solids (i.e. earth), while Plato treats them under the elemental body "water." The subdivisions from the TL are not found elsewhere. See Baltes, Timaios Lokros, 130-35.

[22]Here the term "image" (εἰκών) means model (*Vorbild*) and refers to the idea or ideal world; see TL 88.

[23]That is, "the same proportions and powers" which he had used to establish the world as a whole.

[24]The subject of this sentence could be either nature (it) or the god (he). Since nature is the subject of the preceding sentence, it is probably also the subject of this sentence. One must note, then, that nature is not responsible for the mixing of the division of the human soul but only for its introduction into a body. See Baltes, Timaios Lokros, 142-46.

[25]The literal meaning of σκᾶνος is "tent." However, the term appears four times in the TL (46, 60, 62, 86), and each time it refers to the bodies of living beings. The word first appears with the meaning of "body" in Democritus (Diels, *Frag.* 68B, 37, 57, 187, 223, 270) and is often used with that meaning in Pseudo-pythagorean literature (see H. Thesleff, *The Pythagorean Texts of the Hellenistic Period* [Acta Academiae Aboensis, Humaniora 30.1; Åbo: Åbo Akademi, 1965] 257, Index, σκῆνος).

[26]The terms σπέρμα (semen) and γόνος (seed) are usually synonyms. It is not clear how the two are to be distinguished here. In *Tim.* 73c-d Plato distinguishes the "divine seed" (θεῖον σπέρμα) from other mixtures of seeds. The "divine seed" for Plato means the brain (73c-d) and semen (91a-b) because it bears the immortal part of the soul. The other types of seed bear the mortal parts. That, however, does not seem to be the distinction in this section of the TL. Anton (*De origine libelli*, 279-81), pointing to Diogenes Laertius (7.136, 158) and Eusebius (*Praep. Evang.* 15.20.1) suggests that σπέρμα means the effective element of generation and γόνος is the moisture

in which the σπέρμα is contained. Baltes (Timaios Lokros, 153) suggests that the division refers to fluid from the brain being divided and flowing into either the right testicle (which would engender a male child) or into the left testicle (which would engender a female child). He points to Pliny (*HN* 8.188) and the Hippocratic treatise, *De superfetatione* (31; 8.500). None of the suggestions is really satisfactory.

[27] The sense of touch offers a preliminary notion of "heavy" or "light," but it is reason which really *defines* the meaning of "heavy" or "light" by an object's inclination toward or away from the center of the world.

[28] I have left the word *pneuma* untranslated. When it refers to the substance that fills the nerve canals, it means something other than "breath" or "air" in the usual sense of those words. For appropriate references to the use of the term in ancient medical literature, see Baltes, Timaios Lokros, 171-72.

[29] The meaning of the word λαμπρός (bright) as a primary color is not clear. The same is true of the corresponding section on colors in Plato's *Timaeus* (67c-68d). See Cornford, *Plato's Cosmology*, 277-78.

[30] A basic division of sensible properties is between those properties that draw elemental bodies together and those that push them apart. This division cuts across the various types of sense perception.

[31] The outlets are the nose and the mouth. *Pneuma* here seems to mean "breath" or "air." Yet in TL 65, air (ἀήρ) and *pneuma*, while similar, are distinguished. Then, in TL 67, *pneuma* seems once again to mean something other than simply "breath" or "air." For that reason, I have consistently kept the word *pneuma*.

[32] A cupping-glass is a glass which is heated and then placed over a wound or infection. As the glass cools a partial vacuum is created and the poison sucked from the wound or infection. The ancients thought this was due to the fire consuming the air. Similarly for the ancients, when amber was rubbed, heat was generated and, because of the expansion of the "pores" of the amber, *pneuma* was expelled. To replenish the lost *pneuma*, the nearest object was attracted to it.

[33] That is, substances kindred to the better substances, such as flesh, blood, marrow, etc. *Pneuma*, bile, and phlegm are not bad in themselves but become so when they move into places in the body where they should not be.

Notes to the Translation

[34] Since "them" is feminine in the Greek text, it must refer to "assorted desires and immoderate pleasures" at the end of TL 72.

[35] The author of the TL seems to have in mind the Stoic definition of a passion (πάθος) as an "excessive impulse" (πλεονάζουσα ὁρμή) (SVF 1.50). This viewpoint is also found in Eudorus of Alexandria (*apud* Stobaeus, *Ecl.* 2.44.5).

[36] Different kinds of living creatures are meant to lead different kinds of lives. The author has more than human life in mind, although his interest is focused mainly on the human.

[37] In this context, "music" has a wider meaning and includes rhetoric and science; it is identical with *paideia*. See Plato, *Leg.* 967e, *Rep.* 536d.

[38] It is not clear whether "daimon" refers to the highest part of the individual soul (see *Tim.* 90a-d) or to a divinity independent of man. See Baltes, Timaios Lokros, 138-39.

[39] The Ionian poet is, of course, Homer.

[40] The false stories found in Homer are brought up only when needed, that is, when someone does not follow a healthy way of life voluntarily. However, the unusual punishments described in this paragraph *must* (ἀναγκαίως) be mentioned. The reason seems to be that these latter punishments are *real* and not just a matter of useful but false stories. See *Phd.* 81e-82c; *Tim.* 91d-92c; and Baltes, Timaios Lokros, 243-44.

[41] Once again the term "image" (εἰκών) is used in the sense of model (*Vorbild*), and the genitive which follows is an appositional genitive. See TL 43.

INDEX VERBORUM

A

ἀγαθός   1,7,9
ἀγγέλλω   27
ἀγεμονία   47
ἀγεμών   82,88
ἀγένατος   3
ἀγέννατος   30,88(ἀγέννητος)
ἀγνοέω   19
ἄγνοια   83
ἄγριος   71
ἄγω   7,25,29,60,74,81,83,85
ἀδιάλειπτος   38
ἀδιάλυτος   14
ἀδονά   49,72(2)
ἀεί   3,11,27(αἰεί)
ἀεροπόρος   86
ἀήρ   12,31,32,35(2),36,39,
    40(7),42,58,61,65
Ἀίδας   84
ἀίδιος   4,30
αἰθέριος   24
αἷμα   69(2),81(2)
αἴσθησις   6,48,50,
    51(αἴσθασις)
αἰσθητικός   71
αἰσθητός   2,11(2),59
αἰτία   1,68,70,72,77
αἴτιον   9,10
αἰών   16,30(2)
αἰώνιος   24,88
ἀκήρατος   13
ἀκίνατος   3,4
ἀκμά   66
ἀκοά   51,58,58(ἀκουά)

ἀκοινώνατος   34
ἀκρίβεια   17
ἄκρον   39,40
ἀλαθής   83,85
ἀλγεινός   69
ἅλες   42
ἀληπτικός   81
ἅλιος   26,27(4),28,45
ἀλκά   76
ἀλλάλων   14,15,35(ἀλλήλων)
ἀλλοῖος   74
ἀλλοιόω   7
ἀλλοίωσις   38,69
ἀλλοιωτικός   44
ἄλλος   8(2),9,16(2),17,28,
    29,31,34(2),35,36,45,
    46,59,71(2),84,88
ἀλλότριος   70
ἀλμυρός   56,59
ἄλογος   8,46(2),58,82
ἀμαθής   86
ἀμαθία   71
ἀμαυρός   69
ἀμάχανος   15,62
ἀμενηνός   48
ἀμέρα   27,29(2),76
ἀμερήσιος   29
ἀμέριστος   3,18
ἀμετάβλητος   34
ἄμετρος   72
ἀμιόλιος   33
ἀμιτετράγωνον   33(ἠμι-),33,
    34
ἀμιτρίγωνον   33,35

ἀμοιβαδός 38
ἄμορφος 4
ἀνάγκα 1(2),62
ἀναγκαίως 86
ἀνάγω 1
ἀναδίδωμι 48
ἀνακαλέω 83
ἀνακεράννυμι 72
ἀναλαμβάνω 65
ἀνάλαψις 50
ἀναλογία 6,14,39,63
ἀνάλογος 79
ἀναλύω 70
ἀνάπαλιν 40
ἀναπνοά 63(2),64
ἀνάπνοια 61
ἀνάπτω 50
ἀνάρμοστος 58
ἀνατείνω 64
ἀνατολά 25,27,28,29(2),31
ἀνδρειότας 79
ἀνείργω 85
ἀνεπαίσθητος 48
ἀνθρώπειος 83
ἀνθρώπινος 44,46,87
ἄνθρωπος 51,88
ἀνίατος 69
ἀνισόπλευρος 33
ἄνισος 28,33
ἀνισότας 42
ἀνόητος 86
ἀνορεξία 71
ἀντεπεισάγω 64
ἀντεπιφέρω 64
ἀντικατάγω 62
ἀντικαταλαμβάνω 70

ἀντιλαπτικός 51
ἀντίλαψις 48
ἀντίτυπος 52
ἄνω 54
ἀνώλεθρος 9
ἀόρατος 61
ἀοριστός 7
ἀπακριβόομαι 10
ἀπαναλίσκω 61,62,65
ἀπαραίτητος 84
ἀπαρεγχείρητος 10
ἅπας 8,25,33,37,46,87
ἀπειθής 84
ἀπελαύνω 70
ἀπεργάζομαι 43,44,75,81
ἀπλανής 27,29
ἁπλός 68,73,76
ἀπογέννημα 6,32
ἀποδίδωμι 16,26,29,38
ἀποκαθαίρω 83
ἀποκαθίστημι 49
ἀπόκειμαι 84
ἀπόλαυσις 82
ἀπολείπω 11,13,16,37
ἀπορρέω 61,66
ἀπορροά 66
ἀποτήκω 56
ἀποτομά 31(2)
ἀπόχυμα 47
ἄπρακτος 86
ἁπτός 12,39
ἀργία 77
ἀργός 86
ἄργυρος 42
ἄρδω 60
ἀρεμέω 16

ἀρέμησις 82
ἀρετά 73,76,78,79
Ἄρης 28
ἄρθρον 47
ἀριθμέω 15
ἀριθμός 19,21
ἀριστός 1,9,14(2),88
ἁρμονία 81
ἁρμονικός 19,26
ἁρμός 67
ἁρμόσδω 78
ἄρρην 5
ἀρχά 1,5,18,32,34,47,68, 69,73,78,80
ἀρχαγός 24
ἀστήρ 27,30
ἀστρονομία 27
ἀσυμμέτριος 68
ἄσφαλτον 42
ἀσχημάτιστος 4
ἄτακτος 7(2),58
ἀτρεκής 28
ἀτρεμίζω 82
αὐγά 27,42
αὔξα 14,66
αὐξάνω 70
αὖος 42
αὐτάρκεια 83
αὐταρκής 13
αὐτόματος 7
ἀφά 52(2),53,56,59
ἀφαιρέω 21
ἀφανίζω 27
ἄφθαρτος 9
ἀφίστημι 77
ἀφοράω 10

Ἀφροδίτα 26,27
ἄφρων 46
ἄψυχος 8,63
ἀώς 31

B
βάθος 69
βαρύς 34,35,53,58,72
βάσις 31,33,35(3)
βελτίων 7
βία 1
βίος 78,83
βραδύς 58

Γ
γᾶ 12,31(3),32,34(3),36, 39,40(4),57
γένεσις 38,51,69
γενέτωρ 24,77
γέννα μα 9(2)
γέννασις 32,43
γεννατικός 4
γεννατός 9,10,24,30
γεννάω 3,10,28,30(2), 31(2),32,44,59,68
γένος 34,59,68
γεοειδής 48,57
γεῦσις 56,59
γίγνομαι 1,7(2),10(2),21, 27(2),31,35,48,61,63, 69,74
γλυκύς 56
γλῶττα 56
γνωρίζω 6
γόνος 47
γυμνάσιον 76,80
γυμναστικός 80

γυναικεῖος 86
γωνία 33,34,35(3)

Δ

δαιμόνιος 41
δαίμων 83,87
δαμιουργέω 30,88
δαμιουργός 7
δειλός 86
δεῖμα 84
δεκά 21
δεσμός 39,41
δεύτερος 87
δέχομαι 4(2),7(2)
δέω 21,78,79,80
δήλομαι 7,9(2)
διάγω 17
διαδέχομαι 44
διαδίδωμι 60,67
διάδυσις 56
διαιρετός 24
διαιρέω 19,23,56,59,62
δίαιτα 76,80
διακρίνω 59
διάκρισις 7,56
διαλύω 9
διαμένω 9,31
διανέμω 44
διάσταμα 15
διαστατικός 55
διατίθημι 57,73
διαφθορά 69
διαχέω 59,60
διήκω 58
διικνέομαι 58
διίστημι 57

Δίκα 15
δικαιοσύνη 79
δίοδος 67
διοίκησις 88
διορίζω 57
διπλασία 21,33
δοκέω 55
δόξα 6,83(2)
δριμύς 56,59,69
δρόμος 28,29
δύναμαι 16(2),39,51,59, 74,76
δύναμις 1,18,20,33,41,44, 45,52,68(2),71,80,81(2)
δύο 1,5,18(2),26,29, 33(3),39,40(3)
δυσαισθησία 71
δυσδαίμων 84
δυσθυμία 75
δύσις 25,28,29(2),31
δυσκίνατος 34
δύσμικτος 18
δυσωδής 57
δώδεκα 35
δωδεκάεδρον 35

E

ἔγγιστος 35
ἔγγονος 2,5
ἐγείρω 80
ἐγκελεύω 80
ἐγκέφαλος 47
ἐγκύκλιον 16
ἑδραῖος 34
ἐθίζω 82
ἔθος 76,82
εἶδος 5,32,35,42(2),57,88

Index Verborum

εἰκοσάεδρον 35
εἴκοσι 35
εἴκω 52,85
εἰκών 30,35,43,45,63,88
εἰς 8,18,29,40,41,45,62,68
ἑκατόν 21
ἐκδίδωμι 86
ἐκκαλέω 82
ἐκκρίνω 65
ἐκλείψια 28
ἔκλυτος 72
ἐκμαγεῖον 4
ἐκμελής 58
ἐκπλήσσω 84
ἐκπυρόω 56
ἐκροά 64
ἐκτελέω 26,28(2)
ἐκτυλίσσω 29
ἐκφροσύνα 71
ἔλαιον 42
ἐλάσσων 33
ἐλαχίστος 33(2)
ἕλιξ 29
ἕλκω 61
ἐλλείπω 68
ἐμμελής 58
ἔμμηνος 26
ἔμψυχος 8
ἐναγής 84
ἐνάγω 45
ἐναέριος 60
ἐναλλαγά 40(2)
ἐναντίος 56
ἐναρμόνιος 78
ἐνδιατρίβω 83

ἐνιαύσιος 26
ἐνιαυτός 28,29,30
ἐντοσθίδια 47
ἐντίθημι 83
ἐντρέφω 76
ἔνυδρος 86
ἔξ 21,35
ἐξαγγέλλω 72
ἐξακάτιοι 21
ἐξάκις 35
ἐξάπτω 18,31,69,72
ἐξίστημι 49
ἔοικα 56
ἐπάγω 18,66,77,84
ἐπαινέω 84
ἐπαναφέρω 25
ἐπανορθόω 82
ἐπάρδω 66
ἐπιθυματικός 46
ἐπιθυμία 72(2),82
ἐπικαίριος 70
ἐπιλαμβάνω 16
ἐπιμέλεια 80
ἐπίπεδος 33,40
ἐπίπλαξις 80
ἐπιρρέω 61
ἐπιρροά 60,66
ἐπίρρυτος 45
ἐπιστάμα 6,19,50,83
ἐπίσταμαι 27
ἐπιτρέπω 88
ἐπιφαίνω 61
ἐπιφάνεια 17
ἐπόγδοος 21
ἕπομαι 1,27,84

ἐπόπτης 87
ἑπτά 26
ἔργον 77,80,82
ἐρείδω 31,38
ἔρημος 31
Ἑρμᾶς 26
ἔρως 72
ἑσπέρα 25
ἑσπεριός 27,28
ἕσπερος 27
ἑστία 31
ἑστιάω 21
ἕτερος 18,24,25,26,45,46
εὖ 16,73,78
εὐαισθησία 78,79
εὐδαιμονέω 83
εὐδαιμόνων 83
εὐθυωρία 6
εὐκινατότατος 35
εὐμοιρατέω 45
εὐπείθεια 82
Εὔριπος 64
εὔροια 83
εὐωδής 57
ἐφαμέριος 44
ἐφάρμοσις 15
ἐφέλκω 65
ἔχω 5,16,25,28,33,34(2), 35(2),42,73,78,81
ἑῷος 27(2),28

Z

Ζεύς 28
ζωά 68
ζῷον 11,17,43,44,60,62, 67,78,88
ζωτικός 52

H

ἦθος 76
ἥκω 36
ἤλεκτρον 63,65
ἡμιτετράγωνον 33,34 (ἁμι- 33)
ἧπαρ 46,58
Ἥρα 26

Θ

θάνατος 68
θάτερος 4
θέα 50
θεῖον 42,83
θεός 1,7(2),9(2),20,24, 30,31,39,50,87,88
θεραπεύω 81
θερμός 55,59,74
θερμότας 52,61,68
θῆλυς 5
θηρίον 86
θνάσκω 67
θνατός 17,43,44
θραυστός 42
θρύπτω 76
θυμοειδής 46
θυμός 72,82
θύραζε 64
θυραυλία 76

I

ἰατρικός 80,81
ἰδανικός 30
ἰδέα 2,3,6,7(2),10,86
ἴδιος 28,34
ἱδρύω 31,46,70
ἱερός 27

Index Verborum

ἱκανός 40,76
ἵμερος 72
ἰσόδρομος 26
ἰσοδυναμία 14
ἰσοκρατία 15
ἰσονομία 41
ἰσόπλευρος 33,35
ἴσος 16,33(2),35,41,62
ἰσοσκελής 33
ἰσότας 66
ἵστημι 15,35
ἰσχύς 78,79
'Ιωνικός 84

K

καθαρός 42,81
κάθετος 33
καθίστημι 7
κακέω 70
κακία 72,73,76,77
κακός 73
κάκωσις 69
καλέω 26,33
κάλλος 9,10,78(2),79
κάπρος 86
καρδία 46,66
κάρρων 7,25,70,73
καρτερία 82
κασσίτερος 42
κατασκευάζω 8,39
κάτω 54(2)
κενεός 37
κενός 61,62
κέντρον 54
κένωσις 62
κεράννυμι 18

κεφαλά 46
κῆρ 13,68(κάρ)
κηρός 42
κίνασις 16,18,25,29(3),47, 48(2),58
κινέω 16,25,82
κίρνημι 59
κλῄζω 52
κοιλία 66
κόλασις 80,84(2),86
κορυφά 33
κόσμος 8,11,18,24,25, 30(4),39,41,43,88
κουφός 53,86
κρᾶμα 18(2)
κρᾶσις 74
κρατέω 14(2),39
κράτος 25,81
κρατύς 10(2),39,81
κρέσσων 8,46
κρίνω 52
Κρόνος 28
κρύσταλλος 42
κρύψις 28
κύβος 34
κύκλος 26(2)

Λ

λαβρότας 74
λαγνεία 74
λάγνος 86
λαγχάνω 83
λάθα 71,75
λαμβάνω 14
λαμπρός 59
λέγω 3,20,30,59,66,72,86
λεῖος 17,56(2)

λειότας 52
λεπτομέρεια 36
λεπτομερής 35,55
λευκός 59(2)
λίθος 42
λογικός 8,45,46(2),71,82
λογισμός 6,82
λόγος 1,5,7,15,19(2),26,
   32,41(2),44,51(3),53,58,
   82,84,85
λοιπός 47
Λοκρός 1
λύπα 72
λύσσα 71
λύω 67

M

μακάριος 9
μαλακός 52
μάτηρ 4,5
μαχανάομαι 43
μέγας 33,58,74,76,83
μέγεθος 27
μέζων 33
μεθόριος 66
μείγνυμι 45
μείων 66
μελαγχολία 74
μέλας 59(2)
μέλι 42
μέλος 51
μένω 14
μερίζω 26,44,47
μεριστός 4,18
μέρος 13,14,24,30,33,45,
   46(2),75,78

μέσος 15,16,31,33(2),40,
   53(2),54,58,80
μεσότας 40
μεταβολά 3,7,16,69
μεταξύ 12
μετενδύω 86
μετέωρος 86
μετρέω 30
μετριάζω 56
μιαίφονος 86
μικρός 58
μναμονικός 71
μοῖρα 19,21,29,45,83
μοιράζω 12
μόλυβδος 42
μονάς 21
μονογενής 8
μόνος 16,24,33,34
μορφή 4,18,32,75,86
μυελός 47(2),69
μυριάς 21
μυρίος 59
μωσικός 58,82

N

Νέμεσις 87
νέρτερος 84
νεῦρον 47
νεῦσις 53
νεώτερος 7
νίτρον 42
νοατός 3,11,88 (νοητός 10,
   11)
νοερός 46
νοέω 3,6,66
νόθος 6

Index Verborum

νομεύς 27
νόμος 82,84
νόος 1,6,24,82,83
νόσος 68(2),70,71
νοσώδης 69,81,85
νοτερός 42
νοτίς 61
νύξ 29
νώτιος 47

Ξ

ξυναρμόζω 40
 (cf. συναρμόζω)
ξένος 86
ξηρότας 52,68
ξυμπάς 2 (cf. συμπάς)

Ο

ὄγκος 60
ὀδαξασμός 75
οἶκος 76
οἰστώδης 71
ὀκτάεδρον 35
ὀκτώ 21,34,35
ὀλίγος 58
ὅλος 13,18,23,60
ὅμιλος 27
ὀμιχλώδης 42
ὁμογενής 16,42
ὁμοδρομέω 27
ὅμοιος 16,63
ὁμοίωμα 4
ὁμόλογος 7
ὁμόριος 65
ὄνομα 72
ὄνος 86

ὀνυμαίνω 1,49,52,68
ὀξύς 58,69,74
ὁρατός 12,39,59
ὁράω 7,15(2),24(2)
ὀργά 72
ὄργανον 17,78
ὀργανοποιία 63
ὀρθογώνιος 33(2)
ὀρθός 33(3)
ὄρθρος 27
ὁρίζω 7,53
ὁρίζων 27,31
ὁρμά 74,80
ὁρματικός 71
ὁρμάω 9
ὀρνύω 80
ὄρος 8,11,21,82
ὄρφνα 29,31
ὀσμά 57
ὀστέον 47,69
οὖς 58
οὐσία 4,10,18,20
ὀχετός 60
ὄψις 24,31,50,59,83

Π

παγά 66
παθητικός 71
πάθος 52,70,71,73
παιδεία 80
παιδευτικός 80
πακτός 42(3)
παλαμναῖος 87
πάλιν 15,64
παντελής 11,13,43(-ῶς)
παντοῖος 7,59

παράδειγμα   3,10,11,30(2)
παραδίδωμι   44
παραλλάξ   15
παραφροσύνα   75
πᾶς   4,8,11,13,15,16(3),19, 21,24,25,26,27,31,33, 34,35,36,38,41,66,88
πάσχω   48
πατήρ   5
παχυμερής   55
πάχνα   42
πείθω   82(2)
πέρας   60,73,80
περιγράφω   31
περιδινέω   29
περιέχω   8,11
περικαλύπτω   18
περικατάλαψις   28
περίοδος   26,29,30,87
περιφέρεια   54
περιφορά   38
περίφραγμα   47
πέψις   57
πίπτω   48
πίσσα   42
πλάζω   27,45
πλᾶξις   58
πλείων   66
πλεονάζω   68,73
πλευρά   33,34
πλήρης   37
πνεῦμα   58,60,64,65,67, 70,81(2)
ποθάκω   77
ποθέρπω   29
πόθος   72
ποιέω   8,9(2),20,28,29,35

ποιητάς   84
ποικίλος   59,72(2)
πόλις   76
πολλάκις   27
πολυειδής   59
πολυμερής   35
πολύς   26,27(2),42,58,68, 71
πόνος   80
πόρος   55,56,57,58
ποταγορεύω   4,30
ποταναγκάζω   82
ποταρτάω   17
ποτίγειος   26
ποτιδέομαι   17
ποτιμίσγω   18
ποτίφορος   80
ποτιχρέομαι   39
πρᾶος   82
πρᾶτος   15,21(2),59,68
πρέσβυς   7,20,31,83
προάγω   27
προανατέλλω   27
προγίγνομαι   27
προίημι   51
προκάλυμμα   47
προκρίνω   53
προπέτειος   71
πρότερον   20
προτροπά   80
πτηνός   86
πτοία   75
πῦρ   12,31,32,35(2),36(2), 39,40(4),42,65
πυραμίς   35

Index Verborum   91

Ρ
ῥᾷστος   18
ῥᾴων   21
ῥέος   42
ῥευματίζομαι   75
ῥέω   47,64,70
ῥίζα   31,47,66
ῥίς   64
ῥοπά   31,52
ῥυθμίζω   79
ῥύπτω   56
ῥυσμός   15
ῥυτός   42
ῥύψις   56
ῥωννύω   76,80,81
ῥῶμα   79

Σ
σάρξ   47,56,69(2)
σάψις   57,69
σελάνα   26,45
σικύα   63,65
σκαληνός   33
σκᾶνος   46,60,62,78,86
σκέπα   47
σκληρός   52,84(σκλαρός)
σοφία   45
σοφός   27
σπέρμα   47
σταγών   42
σταδαῖος   34
στενός   57
στερεός   12,40
στερίζω   51
στερρός   57
στοιχεῖον   34,35(4),77

στόμα   64
στόμιον   61
στρυφνός   56(2),59
στυπτηρία   42
συγγενής   51,70,81
σύγκειμαι   35
συγκεράννυμι   19,44
συγκρίνω   18,59
σύγκρισις   56
συλλογίζομαι   21
συμβεβηκός   25
συμμετρία   78
σύμμετρος   58,83
συμπᾶς   1,21 (cf. ξυμπᾶς)
συμπεριδινέομαι   25
συμπιλωτικός   55
συμπληρόω   88
συμπλήρωμα   21
συνάγω   38,57,58,59(2)
συναίτιον   1
συνάπτω   47
συναρμογά   14
συναρμόζω   16 (cf. ξυναρμόζω)
σύνδεσμος   47
συνδέω   39,40
συνδιακρίνω   87
σύνειμι   76
συνεργάζομαι   74
σύνεσις   83
συνέχω   39,60
συνίστημι   19
συντάσσω   9,13,20,26
συντίθημι   14,34,35(2)
σύντονος   72,84
σύρροος   62,81

σύστασις 34,43,67
σφαῖρα 16(2),29,35,54
σφαιροειδής 8
σφόνδυλος 47
σχῆμα 8,15,16(σχᾶμα),56
σῶμα 1,4,7,8,13(2),31,32,
33,34,35,42,46,47,48,
52,55,64(2),65,66,70(2),
72,74,75,78,79(2),80(2),
81,85,86
σωματικός 20
σωτηρία 47
σωφροσύνα 79,82

Τ

τακεδών 69
τάκω 69
τάξις 7,15
τάσσω 7,58,81,82
τάχος 28
ταχύς 58
τέλειος 8(2),11,43(τέλεος)
τελευταῖος 70
τέμνω 33
τέταρτος 59
τέτορες 21,21(τέτταρες),
34,35(τέσσαρες),42,59
τετράγωνον 34(2)
τιθάνα 4
Τίμαιος 1
τιμιώτερος 20
τιμωρία 86
τονόω 80
τόπος 4,15,16,48,70
τραχύτας 52
τρεῖς 2,6(2),15,21,28,33
τρέπω 35
τρέφω 60,80

τριάκοντα 21
τρίβω 38
τρίγωνον 33(3),34,35,42
τριπλασία 21
τριπλάτιος 33(2)
τρίτος 4,5,15,33(2),35
τροπά 7,69(2)
τροφά 47,60,66,67,76

Υ

ὕβρις 86
ὑγεῖα 78,79
ὑγιάζω 85
ὑγιαίνω 75
ὑγιεινός 85
ὑγρός 31(2),42,65,69
ὑγρότας 52,68
ὕδωρ 12,31,32,35(2),36,
39,40(7),42
ὕλα 2,4(3),5,6,7(2),8,
31,32
ὑπάκουος 78
ὕπατος 46
ὑπηρετέω 46
ὑπογίγνομαι 81
ὑπόθεσις 78
ὑποκείμενον 32
ὕστερος 20
ὑπωράνιος 84

Φ

φαίνω 56
φαμί 1,4,51,54
φανερός 28
φαρμακεία 80
φάσια 28(2)
φέρω 60

φθείρω 9
φθίσις 14,66
φθορά 9,38
φιλοσοφία 80,82,83
φλέγμα 69,70
φλεγμαίνω 75
φλέψ 60
φλόγα 42
φόβος 72
φοινικοῦς 59
φορά 25,26,29
φρόνασις 79
φρονέω 48
φυσικός 61
φύσις 1,3,4,8,24,44,46, 49,61,78,80
φύω 51,59
φωνά 58(2)
φῶς 42
φωσφόρος 26,27

Χ
χάλαζα 42
χαλάω 83
χαλεπός 69
χαλκός 42
χειρόκματος 10
χερήων 46
χθόνιος 87
χιλιάς 21
χιών 42
χολά 69,70
χρεία 17
χρέομαι 40
χρόνος 20,26,29,30(4),83
χρυσός 42

χρῴζω 59
χρῶμα 59
χυλόω 56
χυμός 69
χυτός 42
χύω 58
χώρα 4,16,52,70(2)
χωρέω 16,58

Ψ
ψευδής 83,85
ψυχά 18,19,23,44,45,46(2), 58,71,72,76,78,79,80(3), 82,85,86
ψυχρός 55,59
ψυχρότας 52,68

Ω
ὥρα 30
ὡράνιος 24,50
ὡρανός 7,30,31
ὥρος 31